COMMAND AND CONTROL
OF
DISASTER OPERATIONS

by
Walter G. Green III, Ph.D., CEM

Universal Publishers/uPUBLISH.com
USA   2001

COMMAND AND CONTROL OF DISASTER OPERATIONS

Copyright (c) 2001 Walter G. Green III
All rights reserved.

Universal Publishers/uPUBLISH.com
USA   2001

ISBN: 1-58112-659-X

www.upublish.com/books/green3.htm

TABLE OF CONTENTS

TABLES AND FIGURES

PREFACE

This volume was originally written as a study guide to assist individuals preparing for the Certified Crisis Operations Manager examination. It still fulfills that function for certification candidates. The organizational structure reflects the organizational structure of the common body of knowledge, skills, and abilities required for the examination. The focus is the period immediately before disaster onset, the impact, and immediate post impact. This volume does not address other disaster periods or programmatic management of emergency management.

However, the present volume does more than serve as a study guide. I have attempted to codify current best command and control practices in the management of the immediate response to major emergencies and disasters. For students, this is a basic guidebook to the procedures of managing emergency operations. For working emergency managers, this is a reference handbook. And for researchers, this is a partial snapshot in time of practices reported in practitioner literature and observed in field and Emergency Operations Center operations. However, I have critically examined standard approaches and suggested alternatives where, based on my experience and that of others, I believe conventional wisdom to be reasonably subject to challenge.

This volume is dedicated to Cisco, a very busy English Springer Spaniel, who, like many responders, would occasionally translate his experience of life on the high plains into sitting in the window and baying at the moon.

Walter G. Green III
Glen Allen, Virginia, May 2001

1. INTRODUCTION

Major emergencies and disasters are complex events with outcomes that may significantly impact local communities, states and regions, and even nations. Early recognition of the threat, combined with effective management of available resources during and immediately after the event, has the potential to limit death, injury, and physical damage and reduce disruption to political, economic, and social systems.

The Phases of Disaster

Understanding of disaster processes in the United States has focused over the past two decades on a four phase emergency management program cycle used by the Federal Emergency Management Agency[117,256] (see Chapter 8). This cycle makes sense as a programmatic tool for the allocation of staff, funds, and program effort. However, it is questionable whether it accurately describes the life cycle of disasters, and, therefore, whether it is a sound basis for the management of the actual disaster response. If we examine a disaster from the disaster's perspective, rather than from our perspective as victims or responders, the suggested sequence in Table 1-1 might be a more accurate starting point for a discussion of disaster phases.

Not every disaster will exhibit all of these phases. However, the sequence appears to be at least superficially valid. For example, in an earthquake,[17,68,88,102,108] the existing fault and plate structure is the Pre-Existing Condition, which is subjected to the Evolving Condition of increased settlement and infrastructure construction in areas in the potential earthquake zone. The earthquake history of the fault serves

Table 1-1. Suggested Disaster Centric Phases

Phase	Description
Pre-Existing Conditions	Conditions in the natural and built environments at rest before the application of any forces or changes that increase the potential for disaster occurrence or impact.
Evolving Conditions	Changes in the state of nature which gradually increase the level of hazard.
Prodrome	Events which signal the potential for the onset of a disaster, often subtly, and often recognized only in retrospect.
Initial Event	The first event that can be clearly identified as being part of the current disaster.
Contributing Forces	Natural or man-made forces which increase the scope or force of the disaster or change its character.
Impact	The point at which disaster effects are felt by human or animal populations or have a significant ecological effect.
Associated Disaster	An impact of a different type that results from the main impact.
Residual Impacts	Second and subsequent events or phases of the initial event that extend the disaster in space and time.
Restoration of Equilibrium	The gradual dissipation of disaster effects and the return to a non-disaster state of nature.

as a Prodrome[81] to the Initial Event, which may be foreshocks. Failure to adopt adequate building codes becomes a Contributing Force for the event, as the main shock occurs as the Impact. Aftershocks become the Residual Impact, and, as these decrease, Equilibrium is eventually restored. If this earthquake resulted in a dam failure, the resulting flooding would be an Associated Disaster.

Table 1-2 compares this model to the standard four-phase model of Mitigation through Recovery. In trying to understand the actual disaster event from a response perspective, restricting the analysis by incorporating a wide variety of parts of the event into one phase may lead to a misunderstanding of the difficulty of the problem.

Table 1-2. Comparison of Models

Disaster Centric Model	Phases of Emergency Management
Pre-Existing Conditions	Mitigation, Preparedness
Evolving Conditions	Mitigation, Preparedness
Prodrome	Mitigation, Preparedness
Initial Event	Preparedness, Response
Contributing Forces	Mitigation, Preparedness, Response
Impact	Response
Associated Disaster	Response
Residual Impacts	Response
Restoration of Equilibrium	Response, Recovery, Mitigation

The Crisis

In either of these models, there inevitably comes a point at which the situation has become unstable and may take a turn for either the better or the worse. This point is defined as a crisis.[81,93] The crisis is commonly overlooked in emergency management literature. Recently the term has been adopted as a synonym for law enforcement activity in terrorist events. This is clearly a theoretical error, even if its wonderful public relations for the law enforcement community. The crisis in a disaster is a critical point that may occur at any time in the various disaster phases. Recognition of the crisis point and effective action to control the crisis will result in as positive an outcome as is possible.[81] Overlook the crisis, and the disaster will escalate out of control to a far worse outcome.

What are examples of crisis points? On February 4, 1975 the Chinese government directed emergency actions for a major earthquake based on prodromal signals – the next day the earthquake hit. The minimal loss of life (approximately 300) contrasts sharply with the approximately 750,000 killed in 1976 in the Tangshan earthquake.[136,152,269] In 1964 the southern coast of Alaska was devastated by an earthquake;[105] the lack of an effective tsunami warning system at this crisis moment ensured the subsequent associated tsunami that hit Crescent City, California, would be a disaster, resulting in the deaths of 10 and destruction of 150 buildings in the seaside community.[25,269] In 1900 warnings by Isaac Cline, the chief weather observer at Galveston, based on unusual surf and falling barometer (an initial event) were ignored until well past the crisis point at which evacuation of Galveston Island was no longer possible.[128,139] Although these examples are all from natural disasters, there are abundant examples in the literature of man-made disasters as well.[66,81,141,144,180,266]

Managing the Emergency – The Sequence of Actions

In response to the sequence of phases of a disaster, there is a logical progression of actions to be taken by emergency managers. This progression is summarized in Table 1-3.

Table 1-3. Management Actions

Management Phase	Action Described
Communications Watch	A heightened state of awareness based on the potential for emergence of a disaster to allow rapid detection of the actual event – may be based on prodrome or an initial event.
Initial Alert	Initial alerting of response units to allow the start of preparatory actions.
Mobilization	Resources are brought to advanced readiness and prepared for response – resources may be prepositioned.
Initial Response	Initial actions taken to protect the population and to put all command and control resources on the highest state of readiness – such actions as precautionary evacuations, opening of shelters, and activation of emergency operations centers.
Survival Actions	Actions to ride out the impact of the disaster event and to preserve the capability of response agencies to respond.

Management Phase	Action Described
Emergency Response	Actions to protect life, and to a lesser degree property, during the impact of the disaster.
Assessment	Initial actions to determine the extent of the impact and the resulting needs for assistance.
Emergency Recovery	Initial actions taken to reestablish critical lifelines and services and to prevent further avoidable damage.
Recovery	Actions to identify needs, stabilize the impacted community, and reestablish social, governmental, and economic functions.
Reconstruction	Long term actions to restore the community to as close to pre-disaster conditions as practical.

As noted above in the description of disaster phases, not every event will progress through these management phases. A rapid onset disaster, such as a tornado may go directly from the Communications Watch phase (perhaps based on the season) to Survival Actions. It is also important to note that, although it appears there is a logical alignment of disaster and management phases, this may not be so for all events.

The Command and Control Process

Command and Control, and its counterpart Direction and Control, is a process of exercising command and planning and directing operations toward a successful conclusion.[238] As a process, command and control detects and warns, establishes strategy and tactics and develops plans, allocates

Figure 1-1. Command and Control System Model

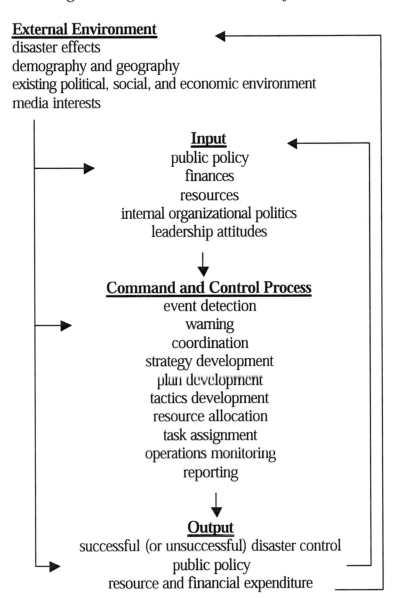

External Environment
disaster effects
demography and geography
existing political, social, and economic environment
media interests

Input
public policy
finances
resources
internal organizational politics
leadership attitudes

Command and Control Process
event detection
warning
coordination
strategy development
plan development
tactics development
resource allocation
task assignment
operations monitoring
reporting

Output
successful (or unsuccessful) disaster control
public policy
resource and financial expenditure

resources, assigns tasks and monitors their completion, and reports results. As a process it is inherently value neutral – when modeled in a systems model (see Figure 1-1 above), the policy inputs determine the social, economic, and political outcomes.

For command and control processes to function efficiently, there must be a command and control system operational in the organization or jurisdiction. This requires at least three components: an organizational structure, management processes, and appropriate facilities from which to operate.[238] The topics in this volume address these three components as shown in Figure 1-2, below.

Figure 1-2. Command and Control Components

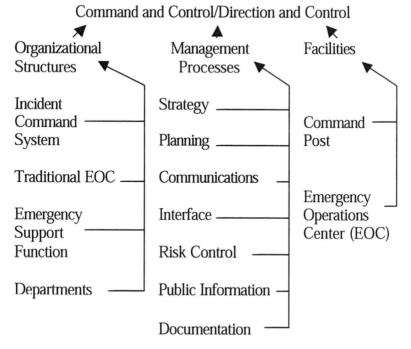

8

2. INCIDENT COMMAND AND MANAGEMENT SYSTEMS

General Applicability

Incident Command Systems and Incident Management Systems provide a standard approach to the management of the site of any large scale or emergency event. There are a wide variety of such systems including those advocated by the National Wildfire Coordinating Group,[165] the National Fire Academy,[257,258] the National Interagency Incident Management System,[230] the Incident Management System Consortium,[157,158] the National Fire Protection Association,[154,155] the American Society for Testing and Materials,[12] and Chief Brunacini[28] of the Phoenix Fire Department. One version, the Standardized Emergency Management System,[30] an outgrowth of FIRESCOPE,[82,250] is mandated in California for operations both at incident sites and in Emergency Operations Centers. Although there are terminology, organizational structure, and focus distinctions between the many systems, I would suggest there are no substantive differences in either structure or outcomes.

Incident Management Structures

General Characteristics

Although not every organization has adopted an incident command or management system, most of those that have, have adopted a system that shares several general characteristics. These include a relatively standard organizational structure, specific names for the various levels of supervision and standard job titles and terminology,[258] and an emphasis on span of control and accountability.[75,198]

Incident Commander

The Incident Commander is the individual assigned overall responsibility for the successful management of the response during an operational period.[163] At the start of an incident, the Incident Commander is normally provided by the organization that has primary legal responsibility for response in a specific area.[33] In general, there should only be one Incident Commander for an incident.[182] Multiple Incident Commanders from participating agencies operating independently diffuse responsibility for overall decision-making, generate coordination problems, confuse everyone, and lead to unsuccessful conclusions.

A number of approaches exist to integrating multiple agencies and jurisdictions into an incident command system when each organization has functional and legal responsibilities for the incident. The most successful approach to unified command under these conditions appears to include:

... integration of Incident Commanders from various agencies into a Unified Command Team
... the integration of staff members from the various agencies into one unified incident command structure
... one consolidated Incident Action Plan
... one Command Post.[168]

In some very large incidents it may be necessary to split the incident into two incidents for management purposes. In this case, each incident will have an Incident Commander and a full Incident Command System. An Area Command Authority may be established to coordinate between the two incidents and to make resource allocations to each incident.[33] In events that become regional in size, a Multi-Agency Coordination System may be activated to coordinate resource and support needs between the incidents and jurisdictions and mutual aid organizations.[164,167]

Command Staff

Command Staff officers provide specific support to the Incident Commander with functions that are not directly involved in service delivery. The key Command Staff functions are:[163]

Public Information Officer - the single media point of contact.

Safety Officer - responsible for identifying safety issues and fixing them - Safety Officers may have the authority to halt operations if needed.

Liaison Officer - point of contact for agency-to-agency issues.

General Staff

The General Staff provides the management of the delivery of emergency response services. Supervision is layered into as many as four levels:[159-162]

Section - four major functional divisions of the staff: Planning, Operations, Logistics, and Finance and Administration. Planning gathers situation data and prepares an overall plan for the next operational period. Operations executes the Incident Action Plan prepared by Planning. Logistics provides Operations the support necessary to carry out the plan. Finance accounts for and manages the costs generated by ongoing operations.

Table 2-1. Incident Command System General Staff[159-162]

Section	Branch	Unit
Planning		Resources Situation Documentation Demobilization Technical Specialists
Operations		Staging Area Manager
	Air Operations	
	Ground Operations	
Logistics	Service	Communications Medical Food
	Support	Supply Facilities Ground Support
Finance and Administration		Time Procurement Compensation/Claims Cost

Branch - a subdivision of Sections based on the number of Divisions, Groups, or Units in the Section -

12

normally only found in the Operations and Logistics Sections. In the Operations Section, Air and Ground Operations may be assigned as separate Branches, or Branches may be used to cover other functions or large geographical areas. Logistics may be split into Support and Service Branches.

Groups and Divisions - these manage resources in the field. Groups are functional in nature, and Divisions are geographic. The term Sector is used interchangeably with Division and Group in some systems.[28,36,48,157]

Units - individual staff functions within the Plans, Logistics, and Finance and Administration Sections are designated as Units.

Typical organizational structure of the General Staff is shown in Table 2-1.

Organization at the Application Level

Field operations are organized on either a functional or geographic basis. Functional organization means that every type of resource that does the same type of thing is grouped together. Thus the organization could include law enforcement or security, medical, utility restoration, fire fighting, evacuation, or similar combinations of resources. Resources assigned functionally are designated by the function and the level of supervision, for example, the Medical Branch or the Security Group. Geographic organization puts resources in the same physical area under the same supervisor.[161] Geographic organizational units are normally lettered (lettering is based on a geographical sectoring of the incident area clockwise) or numbered (numbering is typically used for floors in a building) -- for example A Division or Division 1.[157]

At the resource level, three standard terms are used (see Table 2-2). A single resource is one of anything; for example, a single fire engine is a single resource. A number of resources of the same type organized together under a single leader and answering to a single radio call sign is a Strike Team; for example, five dump trucks might form a Dump Truck Strike Team to move material for sandbagging. Resources of a different type organized together form a Task Force; for example, a police patrol car, fire engine, fire truck, and two ambulances working together under a Fire Battalion Chief would be a Task Force, and the Chief would be a Task Force Leader.[33]

Table 2-2. Field Organization[33]

Branch	Division or Group	Resource
Air	Air Support Group Air Tactical Group	Base Manager single resource
Ground Functional	Group	Task Force Strike Team single resource
Ground Geographic	Division	Task Force Strike Team single resource

Standard Titles for Levels of Supervision

Standard titles are used to identify supervisors at all levels (see Table 2-3).[33,165,169] These provide instant recognition of the level of responsibility even if the function is not familiar.

Table 2-3. Standard Supervisory Titles[33,169]

Title	Position
Commander	Incident Commander
Officer	Command staff officer working for the Incident Commander
Chief	Section supervisor
Director	Branch supervisor
Supervisor	Group or Division supervisor
Leader	Unit, Task Force, or Strike Team supervisor
Manager	individual responsible for a specific limited area such as staging, a cache, a helibase, etc.
Boss	individual resource supervisor

Span of Control

A significant failing of Incident Command Systems has been the relegation of span of control issues to a mantra. The standard phrase is that supervisors should supervise from 3 to 7 people or units, with the ideal number of people to be supervised being 5.[18,48,165] More than 7 workers and the position should be split into two supervisors. Less than 3 and the position should be combined with another one. However, in reality, in complex and high-risk operations a 1 to 2 or 1 to 1 ratio of supervisor to worker may be appropriate. Similarly, with highly competent workers and a low risk situation a higher ratio of 1 to 8 or 1 to 10 may be very reasonable. The important issue is that supervisors have a responsibility to assess work levels and hazards and provide an appropriate level of direction and control for the difficulty and danger of the situation. This level of direction and control may be influenced as much by the reality of the span of communications (the ability to effectively communicate with subordinates) as it is controlled by physical supervision considerations.[48]

Establishing Incident Command

Actions to Establish Incident Command

Incident command should be established by the first emergency unit arriving at an incident site.[28,157,185] For preplanned events, such as major gatherings, at which an incident command system is being used to help manage the event, incident command should be initiated with the start of event set-up. Incident command continues until the last resource departs and the event is closed.

Some systems allow the first arriving unit to immediately start emergency actions and not assume command; command is passed to the next-in unit.[185,257] In theory this allows the first responders to immediately start putting out the fire, controlling the leak, treating patients, etc. without being burdened by setting up a command structure. The danger is that the next unit may be delayed, allowing the event to proceed with no one looking at the big picture. At the same time the second arriving unit cannot supervise or account for the safety of the unit that has passed command until actually on-scene.[157]

As a general rule command cannot be assumed by a unit that, or individual who, is not physically at the location of the event.[158]

The senior, or most qualified, individual in the first arriving unit assumes command by taking the following actions in a measured and orderly way:

(1)   Assume command - announce to the responders on scene assumption of command, make an announcement on the radio frequency allocated to the incident, and advise the dispatch center. Name the incident if not already done so. A typical transmission would be "Unit 7 assuming Building 404 Command."[185]

(2)   Establish a Command Post and do not leave it.[217] Ideally the Command Post should offer a good view of at least two sides of the incident. The Command Post may be as simple as a single spot where the Incident Commander will stand with a clipboard. The important point is that this becomes the focal point for incident management.

(3)   Size-up the situation.[185] Determine what the problem is; how big it is in impact, area covered, people involved; what the hazards are; whether it is getting worse; where resources will stage resources; and the best access.

(4)   Communicate the assessment to the Dispatch Center and request additional resources if needed.

(5)   Establish an initial Incident Action Plan. This is a process of determining what resources are available now, where they can be applied to do the most good, and how added resources will be used as they arrive. It can be communicated orally as orders to subordinates.[48]

(6)   Establish control of incoming resources. Early identification of staging and traffic flow is critical to avoid the scene becoming a parking lot through which resources cannot be maneuvered. Consider having resources staged in a two level process:[48,185]

(a)   Level I Staging. The first resource arriving on the scene proceeds to a location suitable for setting up Command Post and starts incident command. Other initial response resources hold one block short of the location, announce their location, and wait to be directed to their assignments.

(b)   Level II Staging. After the initial response resources have arrived, the Incident Commander establishes a staging location and coordinates with the Dispatch Center to direct second and subsequent resource assignments to that staging area.

Alternative Systems of Establishing Command

At least one author has suggested that there are multiple levels of command, depending on the resources committed to an incident and the complexity of the incident. [157] In this model, shown in Table 2-4, the level of formality of command drives the requirements for a Command Post.[48] The difficulty with this approach lies in the transition from

18

one mode to another, especially in a rapidly developing incident, and the potential for mode confusion.

Table 2-4. An Alternative Model of Command Assumption[48,157]

Mode	Command	Incident Commander	Command Post
Nothing Showing	internal to first unit – not announced	first arriving unit officer	none
Fast Attack	first unit responds to problem – second unit initiates	second arriving unit officer	none
Command	Informal – one unit on scene	unit officer	mobile – moves with Incident Commander
Command	Formal – more than one unit on scene	senior officer	mobile or stationary – located at vehicle
Command	Unified – more than one agency on scene	command team	large stationary in a convenient location

Establishing the Incident Command Staff

As additional resources arrive, the Incident Commander should establish the staff structure to manage the incident. The positions to be established depend on the

19

type of incident, its size, and the available resources. However, the following positions may be needed early in any event:

Operations Section Chief - to take over the direct tactical supervision of operations.

Staging Area Manager - responsible for managing resource staging (with crews aboard and ready to move) or parking (with crews used for other assignments) and the traffic flow of vehicles and equipment.

Logistics Section Chief - to start working supply and support issues for the resources on scene and in route.

Situation Unit Leader - to start tracking the situation and characteristics of the disaster.

Public Information Officer - to start working with the media to provide information needed by the public.

Safety Officer – to manage scene safety concerns.

In general, the incident staff should be located with the Incident Commander at the Command Post.[33,48]

Transferring Command

As the emergency develops it may be necessary to transfer command, either because the situation is large enough that a more experienced Incident Commander is needed, or because of a shift change. Changes of command must be done in person, with the new Incident Commander on scene at the Command Post. The outgoing Incident

Commander should provide a briefing that covers the current status of the incident, the available and requested resources, current strategy, ongoing actions, problems encountered, and hazards. The assuming Incident Commander should formally acknowledge that he or she has assumed command.[48,157,185]

Transfers of command should be limited in number, especially in the early stages of an incident. Each transfer introduces the possibility of information not being fully communicated, and of confusing changes in the style and substance of command. Transfers should be evaluated in the context of whether a transfer of command will improve operations.[28] Unless absolutely necessary, it may be more effective for arriving senior supervisors to serve as back-ups and consultants to the Incident Commander until the situation is stabilized and a smooth transfer can occur. It should be noted that some disagree with this and believe that the senior officer should automatically assume command on arrival on scene.[75]

Early Activation of Incident Command Systems

In some incidents an incident command structure may not be adopted by the agency responsible for managing the response. In other cases, preparation for a major event may suggest early activation of a complete incident command structure for a major function within the incident as a planning tool.[133,166] In either case, an organization or function should be prepared to activate its incident command system and to be incorporated into the disaster incident command system when it is finally activated. For example, the emergency medical services resources on scene could organization and operate as a Medical Group, regardless of whether the responsible agency has established command.[101]

Managing Staff Size

Incidents change in size and in ongoing impacts, both as a result of control efforts and as part of a natural life cycle. During the initial stages of the incident, there is a difficult balance between providing enough staff to effectively manage the response and taking resources from the field where they may be crucial to preventing the situation from deteriorating. Some basic principles apply to this dilemma. First, good management has a significant positive impact on outcomes by making sure that the right resources are applied at the right time and place to get the best solution.[18] Second, a poor start is almost impossible to recover from. And third, the first five to ten minutes of the response will often determine how the entire event will unfold.[74,101]

This means that:

... the initial incident command team, although as small as one or two people, has to do the right things to set up for an expanding incident.[74]

... one of the first calls for any incident that is clearly going to be ongoing for some time should be for additional staff resources. Just throwing more responders at the problem may make it worse if the right staff is not present to coordinate their actions.

... functions flow uphill, and someone has to cover every function in the organizational chart.[36] For example, if the Incident Commander designates two Division Supervisors but does not fill the Operations Section Chief position, he or she becomes responsible for the level of coordination provided by the Operations Section. In a developing incident, these added responsibilities can rapidly overwhelm supervisors.

Staffing needs to be proactive. Ideally, when more resources are requested, additional staff should be in route to provide an adequate number of supervisors and to manage the additional flow of information, actions, and support they will either create or require. A starting point is to look at the span of control, and to order additional supervisors when more resources will push the reasonable limits of span of control.[33] At the same time staffing of the various Units has to be considered; more people and resources mean an increased workload that must be compensated for by more staff.

Staff positions should not simply be filled to generate a complete organization chart.[33] Some functions may not be needed - a Plans Section to deal with a low impact problem that can be resolved in one hour, for example. Some functions may be better performed off site in many incidents. For example, the organization's accounting department may be able to meet all the finance needs after the event from its normal office.

As the incident is controlled, and the number of personnel needed to deal with remaining issues decreases, it is important to start demobilizing resources. Demobilization is a conscious process of releasing resources and staff in an orderly manner according to plan as they are no longer needed. It requires that all personnel and equipment are accounted for, units are intact, and that they are safe to

travel.   Exhausted crews should not be released to drive
some distance to their home organizations - they should be
fed and have a chance to sleep before they leave.   It also
requires that, as staff functions are demobilized, their records
be turned over to the Documentation Unit and any remaining
tasks in progress are handed off to another staff position.   A
smooth demobilization process ensures adequate resources
are retained on site, while allowing those not needed to be
returned to normal service.[207]

Assigning Personnel to Appropriate Staff Duties

Operating in an Incident Command System requires standard
training, experience in the position, and formal qualification
or certification.   Use of a qualification program allows   rapid
identification of individuals who meet the training and
experience requirements and who have demonstrated job
proficiency.   As an example, the National Wildfire
Coordinating Group has a standard Incident Command
System four level training program, based on a series of
modules that progressively cover all of the knowledge
needed to function at the various levels in the Incident
Command System.   For each position, specific requirements
for levels and modules of training, on the job training with
instructor sign-off documenting competency in specific job
element skills, and qualification in subordinate positions
have been established.   Individuals who meet these
requirements are issued a card which documents their
qualifications and which is recognized nationwide.[169]

In a small organization, this level of formality may not seem
to be needed.   However, selection of competent staff and
supervisors is a critical component of the Incident Command
System.   Having a system that trains individuals for the jobs

and identifies who trained people are makes selection easier and provides some defense in case of litigation.

Assign Resources to Organizational Units

As resources report to an evolving major disaster in its early stages, they may be sent directly to an assignment as they check in, whether that check-in is in route by radio or is as they physically arrive. Such initial assignments are made to carry out the initial Incident Action Plan and may be reactive, plugging holes in the initial efforts to contain and control the situation - a scenario that may sound like "great, another ten sandbaggers; go on down the levee a quarter mile; Bob just told me we have a sand boil that looks like trouble, and we need you to help control it; find the guy with the red vest; he will tell you what to do." This has most of the elements of a formal incident command assignment. The Incident Commander has formed an ad hoc sandbagger Strike Team, given them a mission and a location, and told them for whom they are working.

Two days later, the same resources arriving at the scene would probably be directed to a staging area for check in and verification of skills, be designated as Sandbagger Strike Team 7, have a Leader formally assigned, be entered on the incident resource list and Division Assignment sheet, and be dispatched to work for a Division Supervisor. At this point, even if the situation is critical, the needed structure is in place to track the resources and make sure they are being assigned where they are most needed.

In the first case, the Incident Commander is personally involved with resource assignments, the organization is evolving (he may not even know the name of "the guy with the red vest"), and tracking mechanisms are not fully in place (he noted sending 10 more people to the sand boil on his tactical worksheet, but it is probably at the level of an arrow and "+10"). In the second case, the Strike Team was ordered the day before with a report time and location, the planning process has identified the organization and objectives, and accountability systems are fully in place.

This highlights the important steps in assigning resources:

(1)   Have an organizational structure in place. When a resource is assigned to a problem they must have someone to report to - a Division or Group, the Operations Section Chief, or the Incident Commander.

(2)   Know the capability of the resource. Do not assign untrained resources to do anything that requires training (even filling sandbags requires training to make barriers that will not fail). And do not assign trained resources to do tasks for which they are not trained. For example, an organization may have a highly trained data recovery team for its information systems department - that does not say anything about their ability to protect the building against flooding. It sounds obvious, but Incident Commanders do assign resources to do jobs for which they are not trained.

(3)   Have identified tasks to which resources can be assigned. If there are no tasks at the moment, resources are assigned to Staging, positioned with their vehicles and equipment and ready to move as soon as needed.[28] If a task assignment is ready for a resource, the task should be described in terms of the location, the objective, the time, and the measures of success.

(4)   Combine resources as needed to form Strike Teams or Task Forces. These combinations make it easier to supervise the resources, reduce the communications load (as only the Task Force or Strike Team Leader should use the radio), improve accountability and safety, and provide enough people and equipment to have an impact on the problem when they arrive.[33]

(5)   Track assignments to know (a) what resource has been assigned to deal with a specific problem and (b) where that resource is now.[28]

Establish Incident Operational Periods

Most emergency incidents (the typical building fire, medical emergency, mechanical accident, etc.) are controlled in a relatively short time. It is unusual for events to extend for a sufficient period of time to require a second or subsequent shift. However, in large public gatherings, extended community disasters, wildfires, and business disaster recovery, it operations have continued for an extended period of time, days, weeks, and even months.

To manage resources during such extended events, planning should be based on standard operational periods. Each operational period is normally the equivalent of a shift. In most public safety operations these shifts are based on 12 hours, with a day having two operational periods. While 12 hour periods may be appropriate for the immediate response, weeks of 12 on and 12 off has the potential to exhaust work forces, generate bad decision making, and increase the potential for accidents and injury. If distance, travel times, and resources permit, it may be wise to transition to 8 hour operational periods as soon as possible.

27

Operational periods are continuous until the incident is closed. A midnight shift may be staffed only with planners working on the plan for the next day, or by a security staff, or not staffed at all, but conceptually the Incident Commander has decided what to do with that time.

Generally the first operational period starts when the incident starts. However, if the staff that has already worked part of a normal workday, it may be wise to not immediately go to a 12 hour operational period, making their actual workday much longer than the 12 hours. Assess the incident, as there may be a normal cycle of peaks of activity that demand additional resources. It also may be wise to consider how to structure operational periods around normal daily life cycles, especially for staffs that are not normally emergency responders. Consider what works best in the routine work cycle of normal sleep and meal times.

Within the operational period, establishment of a routine cycle of activity is important in large, extended incidents. Each staff section should set standard times for meetings, target times for the production of planning products and orders to resources for scheduled shift activity, updating inventories, briefings, etc.[181] Although there will be immediate requirements for action outside a standard schedule, the schedule allows more efficient use of time and better management of personnel and resources.

Managing Staff and Field Operations

Managing a large-scale incident requires skills in making strategic and tactical choices, coordination, direction, and control.

How a Staff Functions

The power of, and value added, by an incident staff lies in the areas of synergy, problem solving, coordination, information processing, and support for decision-making. Well-run staffs are far more than the sum of the skills of the people present. There is a presumption that anyone assigned to a staff, whether in the Incident Command System or in the Emergency Operations Center, has expertise in his or her field and experience in both exercises and actual emergency operations. By concentrating this talent pool in the Incident Command Post there is a gain from the interchange between experts, their ability to brainstorm and piggyback on ideas, and the ability to interpret the situation in many ways based on the differences in perspectives and experiences. Building an atmosphere in the Command Post that encourages the free flow of ideas and joint problem solving is critical to making the maximum use of this synergy.

Staffs provide the talent pool for problem solving. The Planning Section is the ideal focus for gathering information and identifying the problem. An Operations Section should be able to generate alternatives and test for the best solution. Logistics and Finance Sections can determine what is needed to implement the solution chosen. The Planning and Operations Sections can gather feedback on how effective the solution is proving. To do this requires a regular flow of information between sections, both in regular update briefings (experience suggests that 5 minutes once an hour works well in the early stages of major events) and in on-the-spot discussions as the situation changes.

Coordination is critical to staff effectiveness. Staff members should routinely coordinate their actions with other interested Units in their own Section and laterally with involved Units in other Sections. This horizontal, or lateral, coordination process includes sharing actions in progress, planned actions, reasons for action, information, and support needed from others in order to be successful. Ideally, few major actions should reach the Section Chief level without having support from across the staff at the Unit level. Similarly, solutions should not be presented to the Incident Commander without the involvement of all appropriate Section Chiefs. This requires that information be checked with others and shared with anyone who needs it. Staff members have an obligation to consider the implication of whatever they learn that is new and to make sure that those implications are called to the attention of all interested parties.

Vertical (upward and downward) coordination is much rarer. In the vertical coordination process, the field incident Section Chiefs coordinate with equivalents at the jurisdiction Emergency Operations Center to ensure field solutions can be supported by currently available jurisdiction resources. At the same time, this provides the Emergency Operations Center staff the advance warning needed to effectively work resource requests at their level. Similarly, local jurisdiction functional experts should be talking with their counterpart functional areas in neighboring jurisdictions and in the regional or state Emergency Operations Center.[76]

It is important to note this discussion of lateral and vertical staff coordination is not consistent with some current Incident Command System procedures. The Incident Command System emphasizes strict adherence to chain of command, and commentators discourage lateral coordination

Figure 2-1. Vertical and Lateral Coordination

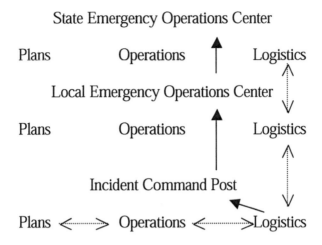

State Emergency Operations Center

Plans   Operations   Logistics

Local Emergency Operations Center

Plans   Operations   Logistics

Incident Command Post

Plans <········> Operations <········>Logistics

Note: Lines with open arrowheads indicate horizontal and vertical staff coordination. Solid lines indicate traditional chain-of-command reporting.

among its components.[218] This centralization of information flow creates bottlenecks and is a significant potential weakness of incident command and management systems when they attempt to operate at prolonged incidents as part of an overall disaster management system.

The Staff exists to support effective decision making by the Incident Commander. Therefore, recommendations for action that reach the Incident Commander must have been coordinated with all staff Sections. At the same time the staff requires regular feedback from the Incident Commander on his or her objectives, intent, and tactics for resolving the incident.

Direction, Command, Control, Coordination, Management

A variety of terms are used to describe the processes used to manage major events. There are differences, sometimes subtle, in the meaning of these terms:

... Command is the legal and organizational authority to direct and control operations in the incident.

... Direction is the process of providing strategy, tactics, instructions, and orders to individuals and units with the expectation they will carry them out in order to meet established objectives.

... Coordination is a sharing process through which organizations exchange information, develop agreed upon courses of action, and synchronize activity to manage an incident.

... Control is the process of monitoring and evaluating on-going operations and applying corrections to increase the probability of achieving the objectives.

... Management is the process of using resources to achieve objectives - often expressed as getting things done through people. Management assumes that you will apply direction, coordination, and control to problem solving.

Managing Incident Resource Communications

Effective communications are critical to the direction and control of emergency operations. As resources are in route to the scene, the Dispatch Center serves as Incident

Commander, making tactical decisions based on available information. The heavy communications load generated by major incidents, especially those involving large areas, impose limitations on the ability of an Incident Commander or a Dispatch Center to manage the communications process. An important technique for incident command and control is deployment of trained communications personnel as Incident Dispatchers to manage tactical communications.[129,132]

Setting Objectives and Measuring Achievement

In any incident it is important for the Incident Commander to set clear objectives toward resolution of the emergency during the current operational period. Objectives are desired outcomes. They translate the overall tactics for the event into measures that can be used to control and direct operations. They are clear statements of the Incident Commander's intent; they do not express how to do a task or what resources should be committed to it. However, all participants should be able to understand the outcome desired by the end of the shift and be able to work toward its accomplishment.[181]

Objectives always should be measurable within the context of the incident.[166] "To provide emergency medical services" describes performing a function, not a measurable objective. The inclusion of a measure of achievement is critical. Without such a measure, it is difficult to determine whether the objective has been met or to take steps to remedy any failures. Treating and transporting one patient or 500 both meet the objective of "provide emergency medical services," but are vastly different levels of effort and outcome.

3. FIELD COMMAND POST ORGANIZATION AND MANAGEMENT

Role of Incident Command Posts

An Incident Command Post provides a central location for the management of operations on-site at every major incident.[28,75] It is the location from which the Incident Commander[119] and the Special and General Staff perform their work. By concentrating the key staff officers in one location, information is more easily gathered, analyzed, and shared; coordinated planning and operations management can occur; and key decision makers can be easily located when a decision is needed.

In a disaster, there may be multiple Incident Command Posts (one for each incident), coordinating strategy, resources, and information with the jurisdiction Emergency Operations Center. In a single incident there should be only one Incident Command Post.[18,33] There is a rare exception to this rule - in a very few cases, the size of an incident or access problems may divide an incident into two major areas which can best be managed by separate Incident Commanders, each with their own Command Post.[33] In this situation the Incident Command Posts have coordination between the two halves of the incident as a primary task.

Command Post Configurations

Folding Command Posts

The New York City Fire Department uses a large folding box as a mobile Command Post.[4.198] This box includes a folding mount for stability. The lid of the box opens to become a magnetic display board used to track resources depicted by magnet-mounted emblems on a large tactical worksheet.[119] Box designs can include communications to provide the Incident Commander a complete workstation. This design is ideal for a Command Post that will have to work in a limited location in close proximity to the incident, including use in inside locations. Set-up is very rapid, allowing the Incident Commander to quickly use the workstation tools to deal with the incident. However, it is designed for use by a limited number of persons (those that can gather around it) and does not include computer capabilities, reference materials, or a variety of communications modes.

Vehicle Plug-Ins

A number of agencies use either agency built or commercially provided plug-ins designed to fit in the rear of a command vehicle, such as a Chevrolet Suburban. These plug-ins typically resemble miniature desks with storage space and drawers for plans and reference materials, forms, and other supplies. When the rear door of the vehicle is opened (or the tailgate dropped), the Command Post is available for work.[130] Tactical worksheets or worksheet boards can be mounted on the tailgate. Communications are provided either by the vehicle radios or by staff officers' handheld radios. This provides the ability to carry more information with the Incident Commander and has the advantage of making the visible command vehicle serve as the indicator of where the Command Post is located. However, working area for staff is still limited, working on a cold or rainy day is uncomfortable, and it is difficult to use the Command Post effectively for interior events.

Command Post Vehicles or Trailers

The third option is the use of a vehicle or trailer as a permanently configured Command Post. The variations of this option range from response vehicles that have built-in space for a Command Post as a module of their construction[7] to vehicles purpose designed as Command Posts,[9,21,130,133] including vehicles as large as tractor trailer combinations.[130,149] The advantages of this approach include enclosed shelter from the environment, the ability to build in work stations designed for specific functions, the arrangement of work stations to maximize staff productivity, the ability to accommodate more staff, and access to a larger communications suite. The disadvantage is that these vehicles are relatively expensive investments for a relatively low use rate. In the case of dual function vehicles, the Command Post and response functions may dictate conflicting requirements for vehicle positioning and the duties assigned to its personnel. Although a number of jurisdictions have converted older vehicles (the retired transit bus,[130,149] schoolbus, or step van[130,193] has been a common choice), a vehicle at the end of its useful life, and not designed for the Command Post function, often means maintenance problems and operational compromises.

Fixed Facilities

At some point in large incidents, the Command Post will have to transition from a rapid response mobile environment to a fixed facility with enough space to accommodate the incident staff.[48] Such transfers are always disruptive -- an early decision to move into a fixed facility is probably a good idea for any event that will last more than one operational period. Such a move involves establishing temporary workstations in a large room, routing communications, setting up displays, etc. The mobile Command Post must continue to operate until the fixed facility is ready; after that a well-equipped mobile Command Post may still be used for specialized roles such as a Communications Center.

Command Post Equipment

With current technology, a mobile or temporary fixed facility Command Post can offer all of the various communications and data management options that a purpose built Emergency Operations Center or Communications Center can offer, subject only to limits of space, budget, and the working environment. The following items should be considered for inclusion in a mobile Command Post:

(1)   Voice (radio and cellular telephone), wireless data, and facsimile communications.[204]   Depending on the function, remote television capability with an on-site camera or the capability to use a digital camera to provide electronic images may be useful. Depending on the incident, hard wire field phones may be desirable.

Amateur Radio and Citizens Band volunteer emergency communications organizations have designed a variety of different approaches to packaging a full suite of communications equipment in limited space. These have included innovative transportable package approaches, such as purpose built[19,121] or fishing tackle box mountings,[32] and detailed emergency station component lists.[34,104,178]

(2)   Access to news and other data sources, including weather radio and commercial weather satellite[216] and

remote on-site weather sensing, broadcast radio, and broadcast television.

(3)   Automated data management capabilities, including laptop computers[204] or personal digital assistants and incident management software to track resources, map the incident site, and assist in decision-making.

(4)   Display capabilities, including tactical worksheets,[204] area maps, status boards, and computer projection.

(5)   Plans and Standard Operating Procedures.

(6)   Technical reference materials appropriate to the expected incident types.

(7)   Capability for shore or generator power with battery back-up to critical systems and battery chargers for all small battery operated systems.[204]

(8)   Ergonomic seating appropriate for individuals working 12 hour shifts.

(9)   Adjustable lighting[119] and environmental controls.

(10)   Interior materials designed to reduce both exterior and interior noise (designs prepared by public safety personnel typically include floors of metal deckplate and whiteboard or formica walls guaranteed to increase the stress generated for the staff by noise by a significant amount).

(11)   Work stations designed to provide adequate workspace for staff (a one foot deep and two foot wide ledge is not adequate workspace).

Functions Within the Command Post

Which Functions?

The Incident Commander should always be located at the Command Post. If, for some reason the Incident Commander needs to move forward for a brief period to actually inspect some portion of the scene, one staff member remaining must be senior enough in experience to manage the staff functions and make decisions within the command post as his or her deputy until the Incident Commander returns to the facility.[18,75] Normally, with modern communications and remote sensing capabilities, this should be a relatively rare event.

Other functions located within the Incident Command Post will depend on:

(1) The size and complexity of the event. The larger the event is and the more difficult and prolonged it will be, the larger the Command Post will need to be to accommodate the required staff. With ten people in the field, the only person in the incident Command Post will be the Incident Commander; with four thousand, the Command Post will need to accommodate a 100 person or more staff on each shift.

(2) The size of the facility. A folding or vehicle tailgate Command Post can accommodate four or five people before they start to interfere with the ability to do work. A purpose built bus would be crowded with a staff of ten on board. If conditions force continued use of smaller mobile facilities as the incident grows, it may be appropriate to

move routine processes away from the Command Post and retain the specialized facility for information gathering and decision making. However, if a large gymnasium is available to work from, the whole staff should probably all be in the gymnasium. In general, staff collocation is the better solution.

(3)   The function the job performs. The Public Information officer needs free access to the Command Post, but will probably do much of his or her work away from it escorting and briefing media personnel. The Safety Officer is probably most productive out in the field. Group and Division Supervisors need to be physically located in the geographic or functional work areas set aside for their resources. The Staging Area Manager should normally be in the Staging Area. On the other hand, the Section Chiefs should be with the Incident Commander, as should some specific units - for example, the Situation and Resources Units in the Plans Section are critical to decision making.[48]

Staffing Patterns

Staffing of the Command Post will be determined by expansion or contraction of the incident staff, based on the level of staff needed to manage the size and complexity of the incident.[28] Early in an incident staffing will be minimal - based on the availability of individuals to do the staff function, and the demands generated by the relatively small number of initial resources. The initial staff may be no more than the crew of the first arriving emergency vehicle. Note that early is a relative concept; in a wilderness search and rescue event, early may be the first six to twelve hours; in a major urban fire, early may be the first ten to twenty minutes. During the height of response efforts, staffing will

40

reach its maximum. This staffing will slowly decrease as resources are demobilized, although release of staff may be slower than that of resources due to the need to complete close down of functions and incident documentation.

The staffing cycle used in the Command Post will normally be based on shifts with shift turnover determined by the end of each operational period. Staffing should be at its maximum during those operational periods with the highest activity or highest threat (in many types of incidents this will be the day shift). However, some functions may require a more rapid staffing cycle than one person per position per operational period. For example, radio communications operators should be rotated regularly during high activity periods to avoid fatigue-based errors. And contingency plans must be in place for relief of even some complete functions in the case of a critical incident (such as death of a responder).

Qualifications for Assignment to Work in a Command Post

In general, individuals assigned to command post duties should be fully qualified in their specific staff duty and experienced in the type and size of incident to which they are responding. Assignment of a newly qualified individual who has only worked on small incidents prior to the "Big One" has the potential for generating overload. Individuals without the appropriate level of experience can be used, but they may require additional attention and coaching from a more experienced staff member of supervisor. Trainees obviously need to work on actual events to gain experience required for qualification; they should be assigned an experienced instructor to supervise and guide them.

Selecting a Location

As a general rule, Command Post location should consider a number of factors that may differ from incident to incident:

(1)   Choose a location a safe distance from the incident and its potential effects based on the cause and by-products. For example, for a chemical release the command post should be located upwind from the incident and outside the Initial Isolation and Protective Action Zones. In a fire, command post location should be beyond the possible spread of the fire or hazardous fire products.[119] And, in a possible terrorist incident, the location should be not only outside the incident impact area, but also outside any probable kill zone for a secondary or tertiary device.

(2)   In smaller incidents, the Command Post should have a good view of the incident site. If possible, it should have a view of the most difficult portion of the incident or of at least two sides of the incident. A good view includes radio line of sight for simplex handheld radio communications or line of sight to a repeater.[28,48,75] In very large, wide area incidents this is not possible – in these cases the command post may be located in the primary base for the incident.[33]

(3)   The location chosen should be capable of being secured to control access.

(4)   If possible, the location should provide some shelter from environmental effects (this may be as simple as setting

up in the shade on a hot day) and access to support services (such as water fountain, shore power, bathroom).

(5)   If the event is not a terrorist event, the Command Post location should be easily identifiable by those who need to know where it is.   If the event is a terrorist event, the Command Post location should not be immediately obvious. Lights, flags, signs, and reflective lighting are very useful in marking the Command Post – it may be good practice to be able to remove or cover these if needed to reduce the signature of the Command Post as a target.

Initial locations for the Command Post can be determined during incident pre-planning for facilities.   For incidents and locations that are not pre-planned, initial decisions of where to set up the Command Post are made in two basic ways. The most common is that the first-in response unit establishes the Command Post location by virtue of assuming command.   Subsequent changes of command may cause the Command Post to move as the event is better understood or as threats develop. However, the fewer moves the better. Each move disrupts command and causes confusion for at least the period of the move – experience suggests that command functions will not resume their normal flow for as much as an hour after a move in some large incidents.

The second approach is to make the selection of the Command Post location while in route. If maps are adequate, and if enough information is known about the event, a site can be selected prior to arrival. This is not unusual in wide area events that extend beyond the ability of the command post to actually provide a view, including search and rescue incidents and wildfires. In these cases,

43

map searches for Command Post locations should include easy road access, identifying areas with sufficient parking for vehicles and equipment, clear ground for establishment of a helispot, etc.

When a location for the Command Post is identified this should be communicated to responding units, much as the staging area is. It is important to understand that if security is a consideration, location of the Command Post should not be broadcast by radio unless encrypted radio is available or a location can be coded through an agency specific grid reference system.

Directing Operations from the Command Post

As the hub of on-site emergency operations, the Incident Commander and Command Post staff must establish a regular flow of operations. This requires the integration of:

(1)   The normal phases of operations in the type of event.

(2)   Jurisdiction and agency Emergency Operations Plans (which should have established basic priorities, philosophy, and how resources will be mobilized).

(3)   The appropriate strategy and tactics based on the plans and the characteristics of this particular event.

(4)   Available resources.

(5)   The current level of threat to life, property, the environment, and the economic and social fabric of the jurisdiction.

The decision making process in the Command Post is similar to that used in any form of emergency decision-making. In this process the Incident Commander and Staff are guided by Emergency Operations Plans, which provide the jurisdiction or agency specific concept of how to respond to emergency events, priorities, and assign specific tasks and authorities to organizations. Standard Operating Procedures provide the routine procedures that can be applied to performance of tasks required to carry out the selected tactics.

The Staff works most effectively in a Command Post environment because the physical proximity offers the maximum opportunity for sharing of information and coordination of actions.

Multiple Command Post Vehicles at a Single Event

In larger jurisdictions or in events that cross jurisdictional boundaries, more than one agency may include a Command Post vehicle as part of its response to a major emergency. In such situations, the Command Post vehicles provide needed communications capabilities and working space for the agency supervisors. However, to achieve a unified Command Post for the event, all the vehicles must be located in physical proximity to each other in a designated Command Post area.[29] Each Command Post thus becomes a management cell in a larger organization. Interoperability is significantly enhanced if each vehicle has the capability to communicate by voice and data with the other vehicles. Military tactical operations centers typically use multiple command and control vehicles to provide sufficient capability to manage large operations,[268] and planning for major emergencies might profitably include mutual aid requests for additional Command Post vehicles.

4. EMERGENCY OPERATIONS CENTER OPERATIONS AND MANAGEMENT

Organizational Structures for the Emergency Operations Center

There are four commonly used models for the organization of Emergency Operations Centers. The model used may result from long established practice, from legal requirements (for example, California requires the use of the Incident Command System as the basis for all emergency management staff organization[46]), or from practices at the State and Federal Government levels.

Traditional Organization

The traditional model for Emergency Operations Center organization, last described in a Federal Emergency Management Agency publication in 1984, is structured around four groups:[245]

(1)   Policy Group – the Policy Group is composed of senior elected and appointed officials.   Its role is to determine policy and organization wide strategy for the management of disaster response.

(2)   Disaster Analysis and Coordination Group – the Disaster Analysis and Coordination Group identifies the impacts of a disaster by collecting and analyzing available data and predicting the effects of a disaster event.

(3)   Operations Group – the Operations Group manages emergency operations by prioritizing, directing, controlling,   and   coordinating   the   use   of   the   available

response resources. The Operations Group is typically composed of representatives from the various emergency services and other major jurisdiction service operations (the jurisdiction's Transportation and Public Works departments are examples).

(4) Resources Group – the Resources Group meets the needs of the Operations Group for operational supplies and coordinates assistance from other governmental, private, or business sources.

This model has the advantage of involving senior executive leadership in the areas in which it is most effective, the setting of overall policy. The Group structure makes good use of departmental representatives as points of contact for their department's normal roles, requiring less training than need for an Incident Command System or Emergency Support Function model. The strong emphasis on disaster analysis is also an asset.

However, this model lacks a dedicated planning staff, a significant shortfall in large and complex disasters. Its structure does not align cleanly with the Incident Command System, although it is much closer than either the Emergency Support Function or Departmental models.

Using the Incident Command System Organization

When the Incident Command System is adopted for use within an Emergency Operations Center, the Incident Command System structure may be adopted in exactly the same form as described for field operations.[209] The senior appointed or elected official present assumes the role of

Incident Commander and individuals are appointed to fill the various General Staff positions. With this structure comes standard Incident Command System terminology and span of control doctrine. The wildland fire model of Incident Command System, with its emphasis on a planning and operations cycle, is particularly applicable to Emergency Operations Center functions.

The advantage to this organizational structure is that it directly parallels what is happening in the field, making interfaces between the staff in the Emergency Operations Center and in the Command Post easy to identify — in either direction, the Logistics Section Chief knows that his or her equivalent is the Logistics Section Chief. Training materials and programs are readily available to help prepare personnel to fit into Incident Command System roles. And flexibility is increased as individuals trained in the system can fill positions either in the field or in the Emergency Operations Center.

The disadvantage is that this structure requires training and practice. In addition, it may place relatively junior staff from one agency in a position of supervising senior staff of another agency, a situation fraught with both seniority and turf issues. It may place people in position to supervise resources that they do not fully understand. And some functions in business or health care may not easily fit in one overall structure. In particular, the Incident Command System has been suggested as a framework for hospital disaster response.[126] The most prominent example of such an adaptation[114] includes answers in its frequently asked questions page that suggest requiring institutions to adopt an emergency structure that differs significantly from their day to day structure is not a simple task.

Organization Using the Emergency Support Functions

The organization of the Federal response into Emergency Support Functions (see Chapter 8) has caused some states to adopt a similar model for organization of their Emergency Operations Centers. If a state organizes in this way, commonality is enhanced if the local jurisdiction Emergency Operations Center is also structured by Emergency Support Functions.[254]

The twelve Emergency Support Functions (ESF) identify the services the Federal Government is prepared to offer in a disaster. In each case a lead agency is designated, and other agencies are assigned to support the lead agency. Put in a local context, ESF-8 is responsible for Health and Medical response; a local ESF-8 organizational structure might represent the emergency medical services agencies, the hospital, the health department, and the mental health agency.

Because the responsibility for leading an Emergency Support Function is assigned to an agency by virtue of the Emergency Operations Plan, this must be well coordinated in advance to avoid ownership issues. The lead agency becomes responsible for assigning a senior staff member for the Emergency Operations Center. This avoids some of the issues associated with the Incident Command System by ensuring that the staff members have at least a conceptual familiarity with the breadth of their area.

However, interface with the Incident Command System is complicated by the need to deal with several

Emergency Support Functions and by the reality that few people understand the Emergency Support Function structure. In addition, the list of Emergency Support Functions does not completely cover local needs. As a result some states have established a list of additional Emergency Support Functions to address law enforcement, animal populations, etc.[99]

Organization by Departments

In some organizations the Emergency Operations Center is staffed by representatives of departments or agencies who come as experts in their department missions and available resources. In this model, the Emergency Operations Center becomes essentially an extended organizational staff meeting. Problems are assigned to departments that would normally work that type of problem in a non-emergency setting.[123] Thus, instead of assigning the task of obtaining more personnel to the Supply Unit (the Incident Command System model), the job is handed to the Human Resources Department.

The advantage of this approach is that it requires minimal training and no specialized knowledge of an Emergency Operations Center management system. The structure used is that used in day-to-day operations, requiring little adjustment in roles and expertise to be successful. Individuals are more likely to have the authority to commit their department's resources. This eliminates much of the stress associated with being asked to perform in an unfamiliar role, while being supervised by or supervising people with whom you are not familiar.

The major disadvantage is that the field Incident Command System must now interface with a number of positions. Instead of contacting a Logistics Section for support issues, the field might now interface with Human Resources, General Services, Utilities, Transportation, and the School Board. A second issue is that this model allows perpetuation of any existing departmental rivalries and turf battles during the emergency.

Personnel and Section Identification

When the Emergency Operations Center is activated workstations should be clearly identified with nameplates for the assigned functions. Although hanging these signs overhead the workstation[209] or mounting them on poles is a common approach, the same effect can be accomplished with normal desk nameplates. These may be color coded by group (traditional organization), section (incident command or management system), emergency support function type, or department type.

Another useful tool for functional identification are mesh vests, color coded by function[147,209] with the position of the wearer clearly marked on the vest. In a large and busy Emergency Operations Center this visual recognition saves time, and helps individuals remember their roles.

The Facility

Physical Layout of the Emergency Operations Center

The physical floor plan of the Emergency Operations Center depends on the space and funding available for the facility. In an ideal situation, Emergency Operations Centers should be self-sustaining, with the capability to continue operations cut off from the outside world for an extended period of time. Such a facility would need space for:[245,254]

      ... a large main operations center room in which the various staff functions are located.
      ... a smaller Policy Group room for use by senior executive staff.
      ... a Communications Center and Message Center.
      ... a Joint Information Center for use by public information officers and the media.
      ... a kitchen with combined dining room and break room.
      ... male and female bathrooms.
      ... male and female dormitories.
      ... a small clinic or first aid post.
      ... heating, ventilation, and air conditioning, air filtration, and back-up battery systems and generator.[147]
      ... storage for food, supplies, and water[140] (ideally including an independent water supply and wastewater system).
      ... security entry control.

Although many Emergency Operations Centers are now located above ground in normal office buildings, consideration should be given to below ground facilities in areas in which tornados are a significant threat. At the end of the Cold War, the need for below ground, hardened facilities with electro-magnetic pulse shielding appeared to have disappeared. However, with increasing concerns about intercontinental ballistic missile threats and nuclear terrorism from emerging third world nuclear powers, as well as the potential for domestic terrorism, abandonment of hardened design philosophies may have been premature.

In reality, few Emergency Operations Centers are able to include all of the above design elements. Many organizations use a conference room as their Emergency Operations Center and configure it on a temporary basis when needed.[140] If disaster activations are a once-a-year to once-a-decade event, an adequate Emergency Operations may be an identified conference room and several large plastic boxes loaded with plans, checklists, and reference materials; needed forms; additional communications devices (handheld radios and telephone sets); and roll up status boards and maps. If activations are more frequent, configuration of a room as a dual use room (conference or training and Emergency Operations Center) with status boards, projection systems, communications plug-ins, etc. in place, may be justified. At the point where activations become more than quarterly, a special purpose Emergency Operations Center may be necessary, especially if events tend to be prolonged.

Location

Emergency Operations Centers are located where they are located for a variety of reasons: the land or building was already owned by the jurisdiction, the department responsible for emergency management hosted the facility in its existing space, convenience to the primary agencies that staff the emergency operations center, relative importance or unimportance of the function to government, etc. In an ideal world, Emergency Operations Centers should be sited:[125,140,245]

...   in a relatively low hazard area (out of flood plains, upwind of hazardous materials facilities, separated from major transportation routes, etc.), especially in comparison to other critical facilities.[147]   In terms of hazard from nuclear attack this may include siting to use available protection from natural features.[213]

...   for easy access, with more than one access and egress route, and with relatively short travel times for key staff.

...   on a site that enhances, rather than limits, communications pathways.

...   with a helispot with clear approaches to allow helicopter operations.

...   with adequate space for parking.

...   and defensible from a security standpoint.

Location of an Emergency Operations Center within an existing building requires some thought. A room on an exterior wall with an expanse of windows may be good for morale,[209] but virtually guarantees that the Center is vulnerable to wind damage or sniping. If operations security is a concern, windows may allow others to determine you have activated your Emergency Operations Center and may even allow sophisticated listening devices to record conversations. A better location would be in a building interior, some distance from the normal entry to the building but close to an emergency exit, and with easy access to bathrooms, vending machines, and the coffee pot (a bathroom at the far end of the building is a major issue when trying to staff a mid shift with one person). If the Emergency Operations Center will depend on radio or cellular communications, it must be located so that radio antenna cable runs are short enough to avoid significant line loss or so that cellular users can actually access a cell from inside the facility.

Internal Arrangement

Within the Emergency Operations Center there are several possible ways to arrange workstations.[245] If you are using a conference room with a conference table, the natural model is to seat people at the table facing each other. This model may be excellent for information sharing and group decision-making. However, it tends to be very crowded as each person tries to deal with phone, laptop, reference manuals, paper forms, etc. Because the focus of activity is the center of the table, status boards on the walls tend to be ignored and not updated — in this model situation displays either should be computer based or placed on lazy susans located along the center line of the table.

Another model uses tables arranged in a U shape with a walkway down the center that can be used for message delivery and other roaming activities.[149,184] This is a good model in terms of facilitation of message flow, but it requires a larger room if you are going to provide adequate room in the center and adequate sized workstations. Like the conference table model, wall mounted status boards tend to be relatively ineffective in this layout.

A third possible model is a classroom arrangement with workstations facing in one direction and placed in rows or clusters. This structure allows like functions to be grouped together more flexibly and located next to more functions they need to coordinate with than does a U or conference table approach. The forward focus of the layout allows displays to be used effectively on the front and side walls and facilitates briefings as a tool for summarizing status and activities.

A fourth model arranges tables in a rectangle within the room, with staff sitting inside the rectangle looking at the walls. This maximizes the use of wall mounted displays and allows space outside the rectangle to be used for message delivery. This model also seems to reduce distractions. Informal conferences can be arranged by all staff simply turning around in their chairs to face inward into the pit. This model reduces the opportunities for coordination, unless an emergency management software on a local area network is in use. A variant places the workstations along the wall with individuals facing the wall as they work.[13,149,184]

A fifth model places workstations in clusters,[13] arranged either functionally or geographically, and facing the main displays. This pattern allows individuals with like duties to work closely together, and may be ideal when the computer and communications systems in a Center allow easy communications between workstations. This model has been used in large Dispatch Centers, such as that of the Los Angeles Fire Department.[72]

A major shortfall in all Emergency Operations Center designs is overcrowding. For effective work, and stress reduction, each person should be allocated a minimum of two by five feet of table space — a good rule of thumb is one individual per folding table or regular desk. If a staff member needs a regular, desk-sized, workstation to do the job in day-to-day operations, the physical size of the work does not undergo miniaturization in an emergency. At most conference and classroom tables this means that two thirds of the seats need to be taken away. The same issue applies to free space around workstations. Locating rows of tables two feet apart guarantees the maximum number of people in a space — it also guarantees that, when the person in the middle gets up for coffee, their progress to the end of the row will disrupt everyone's work. Finally, dense packing the Emergency Operations Center probably will exceed the capabilities of the air handling system, leading into a slow decline into a swamp-like atmosphere of body odor, humidity, and heat that is conducive only to a desire to leave.

Sections within the Emergency Operations Center should be located closest to those other sections with which they have to work most closely. This is a function of the organizational model adopted and the Center's overall policies and concept of operations.[254] In general, operations functions need to be located close to planning and situation analysis functions; logistics functions need to be located next to operations and finance functions; communications needs a direct path to planning and situation and operations. To some degree the use of computer systems reduces the need for face to face contact; however, it does not eliminate it.

Emergency Operations Center Activation Levels

Most jurisdiction Emergency Operations Centers are not staffed for operations on a day-to-day basis. However, as a major emergency or disaster develops, resources will be readied by organizations with a response role. An equivalent increase in the ability of the Emergency Operations Center to support and manage those resources would seem to make sense.

This process is influenced by several factors. Any increase in Emergency Operations Center readiness has associated

57

costs: staff diverted from other work, overtime, facilities no longer available for their normal roles, disruption of normal work routines, etc. Therefore, activation in stages, with an increasing level of staffing in each stage, is attractive because it minimizes costs, while giving the organization a reasonable level of enhancement in the ability to respond keyed to the level of threat.

Activation levels in general use appear to range from five to three in number, based either on time before expected onset of the disaster[38] or on perceived needs for staffing. A composite model of activation levels for an Emergency Operations Center is shown in Table 4-1.

Staffing the Emergency Operations Center

Staffing Requirements

Staffing requirements for an Emergency Operations Center depend on four factors: organizational model adopted, amount of work that can be performed remotely from the Emergency Operations Center, the shift cycle selected, and the size and complexity of the event:

(1) The organizational model defines the number of duty positions that have to be filled to adequately cover all vital functions. For example, the relatively hierarchical Incident Command System will require a larger staff, when fully implemented, than the Departmental model.

(2) The amount of work that can be performed remotely (and the reliability of your communications systems) determines how many people can work normal shifts in their offices, or home offices, or from far away. The

Table 4-1. Emergency Operations Center Activation Levels[39,86]

Level	Activation	Notification To	Staffed By
1	Full Scale	elected officials, jurisdiction agencies, and other Emergency Operations Centers	all jurisdiction agencies
2	Partial	elected officials, jurisdiction agencies, and other Emergency Operations Centers	emergency management agency and key primary response agencies
3	Monitoring	jurisdiction agencies that have response actions as a normal operating responsibility	emergency management agency
4	Basic	emergency services agencies	emergency management and other agency that needs support for a routine emergency
5	Normal Operations	not applicable	not staffed

Note: Routine emergency in this context is an event that can normally be handled by the jurisdiction's emergency services using routine procedures and normal mutual aid without invoking disaster procedures.[111]

more work that can be shifted away, the smaller the Emergency Operations Center staff.

(3)   The shift cycle determines how many people are needed to ensure each duty position is covered. As a minimum for extended operations, one added person should be available for each position; to cover a position on a 12 hour shift cycle, you need three personnel, and for an 8 hour cycle, four personnel. The extra individual provides a back-up in case of illness, accidents, or a higher priority tasking.

(4)   The complexity of the incident determines how many additional individuals must be assigned to meet the workload generated by the response. A short, simple, low impact event might be handled successfully by one person — a major disaster with a national level response might demand 50 to 100 persons on duty. Incident complexity also drives a requirement for more highly trained and more experienced staff.

Staffing requirements may also reflect the rhythm of the incident. Experience suggests that, if the world is ending, the phone will still not ring between about 11:00 pm and 6:00 am — with the exception of emergency responders most people are asleep at this time. This means the night Operations Section may be lightly staffed. On the other hand, the Planning Section may need to be increased to deal with developing the next day's incident action plan.

Conceptually, staffing should reflect the broad range of capabilities of a jurisdiction or organization, with staff members drawn from each major unit to represent needed areas of expertise. In reality, the staff may be drawn from a small group of primarily emergency services organizations.

In some cases, emergency managers have developed volunteer support teams, with membership drawn from public safety agencies, to assist in activation and initial staffing.[96,206,254] This provides a stable team of personnel, effectively increasing the crisis staffing of the emergency management agency without adding full time staff positions.

Staffing Shifts

Emergency Operations Center staffs tend to work in one of two staffing patterns, the work-until-you-drop shift or the 12 hour shift. Neither is ideal for prolonged operations in a major emergency.

The work-until-you-drop model is based on the assumption that each position is one deep and there is no qualified individual who can relieve an individual assigned to do a job. As a result, highly motivated individuals will work for 16, 24, 36, or 48 hours straight without relief. It is important to understand this is often a self-imposed requirement. It is possible to identify individuals who have imposed this requirement on themselves — when the relief comes in they won't leave, or they come back in three or four hours because they know their relief needs help. Second, it is equally important to understand that this is an erroneous perception. Most disaster tasks can be managed by any manager with reasonable common sense. A mediocre decision by a rested individual may be far better than the best decision an exhausted decision maker operating on the edge of fatigue induced physical illness is likely to make.

Twelve hour shifts may be necessary at the start of an emergency until all available personnel can be organized. However, staff members do not normally work 12 hour shifts in government or industry. With commute and turnover times, 12 hours turns into over 14 hours, leaving inadequate time for rest, especially for individuals whose sleep patterns are disturbed by midnight shifts. The result is fatigue, stress, and poor decision-making. As a general rule of thumb, every effort should be made to transition to an eight hour operational period after the first 48 hours of the emergency.

Basic Emergency Operations Center Procedures

Pioritizing Work

One of the most important functions performed in an Emergency Operations Center is the prioritization of information flow and work assignments. During a major emergency hundreds of tasks may compete for a limited number of resources. This requires that Plans, Procedures, and Checklists must provide clear precedences as to which classes of problems require the first and most immediate response. For clarity and consistency, the same system should be used for information flow, task assignment, and message traffic. The considerations shown in Table 4-2 provide one approach to managing priorities.

Types of Work Performed in the Emergency Operations Center

Emergency Operations Centers perform a wide variety of work during an emergency event.[125,245]

Determination of Policy and Strategy

As a seat of emergency government in crises, the Emergency Operations Center provides the location for

Table 4-2. Work Priorities

Precedence	Color Code	Task
EMERGENCY	Red	rescue of people or animals in danger
EMERGENCY	Red	protection and reestablishment of key utilities or lifelines including communications
EMERGENCY	Red	protection of critical data
EMERGENCY	Red	population protection and mass care actions including evacuation, sheltering, mass feeding, and water supply
EMERGENCY	Red	prevention of significant long term environmental damage resulting from hazardous releases
EMERGENCY	Red	protection of public health
EMERGENCY	Red	logistics and support actions to achieve Red tasks
PRIORITY	Yellow	recovery of human remains
PRIORITY	Yellow	protection and salvage of property
PRIORITY	Yellow	other environmental protection
PRIORITY	Yellow	reuniting of families
PRIORITY	Yellow	logistics and support actions to achieve Yellow tasks
ROUTINE	Green	all other tasks

determination of jurisdiction and organizational policy[184] and the setting of basic incident strategy.[83,245]

Incident Situation Information Gathering, Processing, Evaluation, and Assessment

A primary function is the gathering of data on the characteristics and impacts of the disaster from a wide variety of sources. These sources include reports from the organization's own personnel, damage assessment teams, the general public, news media reports, staff of other organizations and agencies, information from utilities, governmental agency situation reports, weather reports and forecast data, and Internet sources. This information must then be fused into a single picture decision makers in the Emergency Operations Center can use to make reasonable choices on how to best use their limited resources. This process of situation assessment requires that the staff make informed judgments about the credibility of often contradictory data and reasonable extrapolations on how the disaster will develop.

Information Distribution

Emergency Operations Centers perform an important role for organizations as disseminators of current information on the situation, the level of response, and projected actions. The most common product used for information dissemination is the Situation Report (or SITREP), a report published at regular intervals by the Emergency Operations Center and typically distributed vertically and laterally to other organizations.[85,223,226,261]

Requests for Assistance, Resource Allocation, and Mission Tasking

Requests for assistance in an emergency are routed to the Emergency Operations Center from a variety of sources, both official and unofficial. Staff in the Emergency Operations Center must confirm the need for assistance and assign it a precedence in the list of tasks to be performed. This may require coordination with other organizations to ensure that they are not already working on the need in order to prevent a duplication of effort. Based on validation of the task, resources must be allocated using the criteria of task size and difficulty, time requirements, hazards to the responders, and level of effort that will be needed for successful completion. Bigger, more complex tasks that require rapid completion under hazardous conditions will tend to require large amounts of resources. Once a resource is allocated, the Emergency Operations Center assigns a mission to that resource using a standard mission order format that should provide all the information needed.

Mission Monitoring

When allocated resources depart on a mission, the Emergency Operations Center should monitor their progress on a regular basis. There are two reasons for this. First, if the work is not going as quickly as expected, more resources may be needed, or the situation may be so hazardous that the resources need to be recalled. If work is moving faster than expected, the resource may be available for reassignment sooner. Second, regular monitoring performs an accountability function for safety. If a resource cannot be contacted for a regular accountability check, the operations staff should take immediate action to locate, and, if necessary, rescue the missing unit. In addition, information received in reports from the field will help build the understanding of the overall situation.

Reporting

An Emergency Operations Center performs a variety of important reporting roles. The Situation Report has already been mentioned as one tool for this. However, a regular flow of more detailed information upchannel to the next level of government is also important to help that level understand the full impact of the disaster, probable needs for assistance, logistics concerns, etc. Even a short message that certain services are fully functional will allow early reallocation of effort to meet other needs.

Activation and Deactivation Procedures for the Emergency Operations Center

Activation

Activation of the Emergency Operations Center should be considered whenever an emergency incident is in progress that could benefit from application of the Emergency Operations Center's coordination capabilities. There is little reason to activate the Emergency Operations Center for the single-family house fire, the one car-one patient-one telephone pole accident, or the employee heart attack. These events are routine emergencies that can be easily handled by application of the Incident Command System on the site. Depending on the size and resources of the organization, even large incidents may be handled by the Incident Command System with no need for Emergency Operations Center activation.

However, when any of the following considerations are present in an incident, activation of the Emergency Operations Center should be considered:

... potential for the event to be prolonged beyond one operational period.
... resource needs will extend beyond normal mutual aid requirements.
... a large area is involved.
... deaths or significant numbers of major injuries are involved.
... participation will be required from multiple departments, including those without a normal emergency response role.
... increased public, media, or regulatory interest in the event.
... there is significant property damage.
... activation is required by law, either for the event type or as a precondition to assistance from higher levels of government.

The activation process should be established in a standard operating procedure and provide for an orderly sequence of steps to achieve full operational capability. Among those steps may be:

(1)   Immediately open the incident log and record all subsequent actions.

(2)   Immediately open the checklist for the type of emergency, and start carrying out the actions it describes.

(3)　　Alert the Emergency Operations Center staff and provide them a time to report to duty. The use of an alerting roster using a callout tree (one individual calls two, who call others, etc.) will speed that process.

(4)　　Physically open the door, turn on the lights, and put up the sign that says "Emergency Operations Center Activated."　If this is a shared facility, contact the keeper of the schedule for the room and advise that the Emergency Operations Center is being activated so that other users can reschedule their space needs.　If you don't have permission to seize the use of the facility whenever needed, contact the individual who does and request permission to open the facility.

(5)　　Set-up workstations using supplies stored in the room or from your Emergency Operations Center kit.

(6)　　As soon as operations can be supported with a facility, supplies, and staff, start monitoring the situation, and contact all participants to advise that the Emergency Operations Center is operational.　The point to do this may be as soon as the door is open and a phone plugged in, or it may be when a nearly full staff is present — it depends on the incident, who is handling the incident now, what they need and how quickly, and the staff's own comfort level.

(7)　　Assign arriving staff to duties.

(8)　　Determine who can be released now to go home and sleep, so they can come back as the night shift. Managing the transition into 24 hour coverage with fresh staffing is critical.[214]

(9)   Brief the staff on what is known about the incident, what needs to be determined, and what initial actions must be taken.

(10)   Advise executive leadership of the activation, the status of the incident, and what steps are being taken to control it.

(11)   Start carrying out the Emergency Operations Plan.

Deactivation

When is an Emergency Operations Center deactivated? The simple answer is when it is no longer needed. In general, Emergency Operations Centers are most productive when conditions require centrally coordinated and constant interaction among a variety of functions for the effective protection of the business or community. When that same level of protection can be assured through the normal coordination of an organization's staff from their regular offices using day-to-day procedures, the Emergency Operations Center is no longer needed. In practice, deactivation is generally a gradual process during which functions that are no longer needed are released. When 24 hour food service is no longer needed for the staff, the food service representative can be released. When all the shelters are closed Mass Care can go home. When the data center is restored or is working well from a hot site, the information technology representative may be able to go to an on-call status.

If possible, deactivation should be a planned process. Being able to identify in advance when services are expected to be restored to near normal and functions can be released may result in significant savings of personnel and resources costs. If deactivation is included in the daily Incident Action Plan, supervisors can project staffing and service levels more effectively.

Deactivation is not as simple as just saying "go home." There is a series of significant events that should be included:

(1)   Ensure people from functions that are deactivated are safe to travel. Exhausted staff members may need to be fed and allowed to sleep before you allow them to drive home.

(2)   Collect all documentation and file it in the disaster record. Any documentation that leaves the site will never be recovered. If necessary, make copies for an individual who needs them for department records.

(3)   Conduct an informal outbriefing with each section released, and schedule a more formal after action review. Capturing lessons learned is critical to future successful operations.

(4)   Identify anything that needs to be fixed. People remember forms that are clumsy to use, software glitches, keyboards where a key sticks, missing pages from a reference manual, etc. when they are in the room. By the time they get to their cars, the problems are lost.

(5)   Clean the workstations and restock forms, office supplies, etc. The Emergency Operations Center may be activated tomorrow for an unexpected event.

(6)   Before people walk out the door, make certain they have turned in all information needed for payroll, travel pay, or reimbursement purposes. It will take months to get it back if not.

Evacuation and Relocation of the Emergency Operations Center

In some cases it may be necessary to abandon the Emergency Operations Center, evacuate personnel and critical materials, and relocate in another facility. Such a course of action is never desirable, as it completely disrupts ongoing operations, causes great confusion among outside agencies as to how to contact the Center, and creates opportunities for communications breakdowns. However, any jurisdiction or business may be forced to conduct such a relocation during a disaster, and procedures and training should be in place to make the process as smooth as possible.

Two Evacuation Cases

All Emergency Operations Center relocations can be described as falling into two types, an orderly relocation or an emergency evacuation.

Orderly Relocation

In an orderly relocation sufficient time exists to move personnel to new facilities in an orderly, two stage movement. Such an event may be caused by a slowly intensifying disaster in which it is possible to identify when the current Emergency Operations Center will become untenable. Flooding provides one such scenario. Another possible case is a situation in which the Emergency Operations Center has been damaged by disaster effects and is not habitable over the long term.

The first stage of the orderly relocation process is sending an advance team to the new facility to make it ready to receive the full staff. The advance team sets up workstations, arranges for communications and, when ready, takes over the Emergency Operations Center function. When this happens, the remaining staff gathers their materials, turns off the lights, closes the door, and completes the move.

This approach has several benefits. It ensures that some Emergency Operations Center capability remains in place at all times. It allows the transition to be orderly, and it provides maximum time for alerting all participating agencies and organizations to the move. However, it does take time to complete, requires significant transportation resources, and presumes a fairly permissive disaster environment.

Emergency Evacuation

An emergency evacuation is a highly undesirable event, but it is far better than endangering the Emergency Operations Center staff by attempting to remain in an untenable facility. Emergency evacuations are appropriate when the rapid onset of disaster effects is combined with significant damage to the center. Fire, a tornado, flash flooding, an earthquake, or a hazardous materials release might all force an emergency evacuation.

In an emergency evacuation everyone leaves, as quickly as possible. Depending on the situation, it may be possible to salvage some materials from the Emergency Operations Center. When personnel are safely accounted for, relocation to a new facility starts.

Evacuation Procedure

Once a relocation or evacuation is in progress, the staff of the Emergency Operations Center should consider certain basic priorities:

(1)   If you have to, get the people out and nothing else.

(2)   If slightly more time is available, the staff should carry out key documents, including first the log, then cellular phones and lap top computers.

(3)   With more time, standard fly-away kits, prepared in the case relocation is necessary, should be carried out.

(4)   If there is time for phone calls, time should be spent notifying key players that a move is in progress.

(5)   Next should come remaining operations and reference manuals.

(6)   And if there is plenty of time bring wall displays, radios, desktop computers, telephones, and anything except furniture.

Reconstitution

Reconstitution of the Emergency Operations Center requires availability of an alternate site. In an ideal situation, the organization should have an Alternate Emergency Operations Center identified and have a minimum set of furniture, telephone (and antennas and cables for radio) communications, and supplies already in place.[147] More likely, a space will be identified, but no preparations will have been made (in some cases even the normal user of the space is not notified of its intended use). In the worst case, the Emergency Operations Center staff will be standing in the parking lot in the rain trying to decide who has a conference room that can be used.

When the staff has reached the location chosen for the reconstituted Emergency Operations Center, actions must be taken to start an orderly set-up process. These may include (in rough order of priority):

(1) Account for all personnel.

(2) Assign key personnel to oversee the critical reconstitution functions: first the reestablishment of the incident log, then communicating new location and contact information to outside agencies, and finally facility set-up.

(3) Determine what space can be used and what furniture is available.

(4) Set a basic design for how furniture and space will be used.

(5)   Set up a minimal Communications Unit and a minimal Operations Section first to handle immediate needs. Follow shortly with the Plans Section or Analysis Group.

(6)   Determine what has happened while the Emergency Operations Center was in transit, and reestablish a good situation picture.

(7)   If space is constrained, consider each staff position carefully and determine if the function represented can be effectively performed elsewhere.

(8)   Make sure the next shift knows where to report.

(9)   Complete the set-up of the physical facility and resume full operations.

The Alternate Emergency Operations Center

A wide range of events may make it impossible to operate from the primary Emergency Operations Center. Although it is possible to establish an improvised Emergency Operations Center in an available space, it is preferable if an Alternate Emergency Operations Center is established in advance.[245]

It would be ideal if the Alternate is as well equipped as a fully capable primary Emergency Operations Center; however, budget and facility constraints make it likely that an Alternate Emergency Operations Center will be austere. Even an austere Alternate can provide a significant capability if the following guidelines are adhered to:

(1)   Separate the Alternate from the primary, and locate it well out of any hazard areas that may impact the primary Emergency Operations Center.[147]   If the hazard puts the primary out of action, the same event should not also destroy the Alternate.   Convenience may require the primary to be located downtown in City Hall – if so, the Alternate Emergency Operations Center should be as far from downtown as practical.

(2)   Coordinate for space use.   Make certain the agency that owns the conference room intended for use understands the need, the relationship is in writing, and the continued suitability of the facility is reviewed annually in a joint inspection.   It is also important to confirm that both organizations can share the space in an actual major event – the room planned for the Alternate Emergency Operations Center may also be where the other agency normally would brief and stage their emergency crews.

(3)   Store one complete Emergency Operations Center kit with current plans, standard operating procedures, checklists, forms, maps, etc., at the Alternate.   Inspect and update this regularly.   Include a diagram that shows all components and tells users how to set up the space for use.

(4)   Include the Alternate in the Emergency Operations Plan, along with directions on how to reach it and how to gain access outside of normal work hours.

(5)   Include the Alternate in each exercise so the staff will know the location and understand how to operate in it. Options include emergency relocation in one exercise, running the exercise from the Alternate in another, using the Alternate as the exercise control center in a third, etc.

Subordinate Operations Centers

In some jurisdictions, individual departments or agencies may activate and staff internal Operations Centers in major emergencies. Functions performed by such Operations Centers may include:[123,204]

(1)   Accounting for and calling in off-duty personnel.

(2)   Coordination of agency specific mutual aid requests with normal mutual aid partners.

(3)   Activation of overhead teams to supplement the staffing for the on-scene Incident Command System.

(4)   Managing resource distribution to ensure coverage of areas not impacted by the current emergency event.

If such an Operations Center is activated, it is critical that its role is well defined and coordinated with the roles of the on-scene Incident Command System, the Dispatch Center, and the Emergency Operations Center. Failure to communicate could result in multiple and conflicting taskings for resources, duplicate mutual aid requests, and breakdown of situation information flow.

Virtual Emergency Operations Centers

Modern computers systems and the Internet permit effective e-emergency management, including the remote performance of Emergency Operations Center functions. This may be as simple as allowing organization offices to communicate key data electronically and perform emergency work on-line.[208] At a more sophisticated level, the capability has been demonstrated to actually operate an Emergency Operations Center completely on-line using standard procedures by The Virtual Emergency Operations Center.[95,98]

Supplying The Emergency Operations Center

Sustained operations during disaster conditions use a large amount of a wide variety of supplies. Regardless of the organizational model and physical layout selected for the Emergency Operations Center, this reality requires assignment of space for storage and designation of an individual as supply manager.[140]

The support philosophy used by Emergency Operations Centers varies from no capability for self-sufficiency to the ability to operate for prolonged periods without external support.[100] The standard advocated by emergency management for general population preparedness is 72 hours. This same time period would seem to be an acceptable minimum standard for Emergency Operations Centers.[140,213] The following supply list may serve as a starting point toward achieving a 72 hour capability:[140,147,213]

(1)   Emergency generator sufficient in capacity to power all critical systems with fuel for sustained operation and all appropriate cabling and adapters.

(2)   Internal water supply sufficient to provide water for drinking, food preparation, and showers for all staff. A minimum water quantity for planning for austere conditions is 4.6 gallons per person per day.[77]

(3)   Food supply for three meals per person per day.

(4)   Sanitation kits, either individual or group, to replace toilets if the sewer system is inoperable.

(5)   Cots and blankets sufficient for one third of staff to be sleeping at any time.

(6)   Three days supply of all forms used, along with staplers, pens, pencils, markers, erasers, tape, file folders, and flip charts.

(7)   Copier supplies including paper and toner.

(8)   Batteries and battery chargers for all battery operated systems.

(9)   Back-up radios for critical radio systems, including power supplies and cables, all connectors, sufficient coaxial cable to run from the Emergency Operations Center to outside, and antennas that can be easily erected to provide emergency communications.

(10)   Battery operated lanterns and flashlights.

(11)   Laptop computers with current databases, backup batteries, portable printers, and all cabling and connectors, sufficient in number to meet critical data automation needs.

5. USE OF PLANS AND STANDARD OPERATING PROCEDURES

Types of Guidance Documents

Organizations use a wide variety of planning and guidance documents to direct emergency operations. The specific format of similar documents from a variety of organizations will show significant differences in organization and content. Planning documents are most effective if they are developed by the using organization (as opposed to copying boilerplate or depending on a consultant product) and if they represent the policy guidance of senior leadership and the technical expertise in emergency processes of the operational staff.[65]

Plans and procedures are vital to the successful management of major emergencies. Because they are prepared in advance, they can incorporate a wide range of input to ensure every level is addressed from task performance to policy. These documents identify the common routine actions needed to manage any type of event,[81] and free decision maker time during the actual event to focus on resolving the excursions from the routine. And plans and procedures clearly document that managers have thought through the process and developed the best approach that a reasonable person could develop (assuming that a planning process has been used rather than filling in the blanks in a model plan[80]).

Emergency Operations Plan

Emergency Operations Plans are jurisdiction level (village, town, city, county, or state) plans for the management of response to disasters. In general, Plans are formally adopted by jurisdictions= governing bodies and thus

80

have the force of law. An Emergency Operations Plan is a basic public policy document that articulates the priorities of government in the coordination of governmental, voluntary, and private resource response to protect life, property, the environment, and the social fabric of a jurisdiction.[65,256]

The size and complexity of an Emergency Operations Plan varies based on the size of a jurisdiction and the nature of the hazards that may threaten it. A town or village level plan might be contained in a single two inch, three ring binder. At the state level the Emergency Operations Plan will commonly be multiple volumes in size. The standard contents of an Emergency Operations Plan include:[65,246,248]

(1) Basic Plan: The Basic Plan establishes general policy and guidance for emergency response, defines a general concept of how government will respond, and assigns responsibilities to agencies and organizations. Standard sections of the Basic Plan include:

(a) Introduction: The Plan's Introduction includes a wide variety of information needed to understand the document. Critical parts include a promulgation document, typically signed by the chief elected officer of the jurisdiction, which establishes the legitimacy of the Plan, and a hazards analysis that provides a summary of the threats faced by the community. Instructions are typically included for distribution of the plan. And a record of changes page, if kept up to date, allows the reader to immediately establish whether or not a single copy of the document is current.

(b) Purpose of the Emergency Operations Plan: This is a general explanation of the reason for the existence of the Emergency Operations Plan and the types of situations for which it will be implemented.

(c)   Situation   and   Assumptions: The Situation section provides a description of the operational environment government can expect to face when the Plan is activated.   The Assumptions section provides the basic assumptions that the Plan is based on; if those assumptions are not valid, then the Plan may not be valid.

(d)   Concept of Operations: The Concept of Operations establishes basic policies on how the jurisdiction will respond to a disaster. It provides guidance on the priorities government assigns to specific actions to prepare for and respond to a disaster, and establishes a framework for how various agencies are supposed to work together to protect the community.   Properly written, this section gives the reader a clear overview of how, when, where, and with what resources government will respond.

(e)   Organization   and   Assignment   of Responsibilities: This section establishes the organizational structure of government in an emergency, defines which officials and organizations have responsibility for direction and coordination of emergency response, and assigns specific responsibilities to all agencies included in the Plan. Responsibility assignment is usually very specific in terms of the actions an organization is expected to perform.

(f)   Administration and Logistics: The Administration and Logistics section defines how emergency operations will be supported.   This includes how records will be maintained, budgetary issues addressed, needed supplies will be procured and distributed, etc.

(g)   Plan Development and Maintenance: This section describes the procedure used to coordinate and adopt the Plan, provides the procedures for revisions, and indicates how the Plan will be tested and evaluated.

(h)   Authorities and References: This provides the legal authorities for the Plan and its adoption and usually includes reference to specific statutory provisions. The section also provides a list of references that assist the reader in understanding the Plan.

(i)   Definition of Terms: A glossary defines terms unfamiliar to non-specialist readers and also provides emergency specific meanings of common terms.

(2)   Functional Annexes: Functional Annexes address the procedures for performing specific functions in every type of emergency.   Examples might include sheltering, emergency medical services, search and rescue, evacuation, and similar functions that will work essentially the same way regardless of the type of emergency.

(3)   Hazard Specific Appendices: Hazard specific Appendices address the procedures and considerations that apply to a given type of emergency event.   Appendices are normally prepared for the major types of hazards that threaten a community, such as hazardous materials, terrorism, hurricanes, flooding, etc.

Business Continuity Plan

Business Continuity Plans define how a business will survive the impact of a disaster and restore its ability to operate **B** in some industries Business Continuity Plans address how to continue to operate through the impact of a disaster without disruption of service. The focus of a governmental Emergency Operations Plan is the protection of the entire community; the Business Continuity Plan focuses on protection of the assets and profitability of a single entity in the community. One is outward looking, the other inward focused (although it must include attention to supply chains, utilities, and services issues outside the corporation). Although business continuity planning is not commonly done by government agencies, any emergency response organization should have a plan for how it will continue to operate as it is hit by the effects of a disaster.

The content of Business Continuity Plans is far more diverse than that of Emergency Operations Plans. Because of the focus of such plans on specific business processes and the disruption impact of even small events, plans may be prepared at the individual facility or even departmental level. Even when plans are prepared using standard commercial continuity planning software, there may be great variability in the contents and organization of Business Continuity Plans. Paragraphs or guidance that may be addressed in a Business Continuity Plan include:[148,183,222,272]

(1) Introduction: This section establishes the purpose, scope, and authority of the plan.

(2) Hazard and Business Impact Analysis: This identifies what hazards the company may be exposed to and what the expected impact of those hazards will be on company operations at the location and work unit for which the plan is prepared. This should include both hazards to

which the entire community is exposed and specific hazards unique to the business or the facility.

(3)   Concept of Operations: This provides a general policy statement of how the company will respond and what priorities are during each phase of the emergency. Senior company leadership involvement in establishment of this policy is critical because the concept should guide how all employees will work to protect the company.

(4)   Protection Actions in Place: A review of the current actions and measures the company has taken to protect its operations may be useful.   Ideally these actions should already address many of the potential hazards and impacts.   This section is important because many readers will not be aware of the variety of systems already in place.

(5)   Protection Priorities: This section establishes what the organization needs to protect and which protection actions should be taken in what order as the emergency develops to maintain operations or permit rapid recovery.   It should assign specific responsibilities to individual offices for action.

(6)   Notification: The   notification   section addresses how to alert anyone who needs to know that a problem is developing.   This includes people and organizations within the facility and also external notifications, and should establish a logical sequence of who should be told in what order.

(7)   Organization: This section addresses chain of command and continuity of leadership issues, who is authorized to declare a disaster and implement the plan, who

is authorized to make decisions at specific levels of response, where an Emergency Operations Center or Command Post will be established and who will be in it, what types of emergency teams will be established, etc.

(8)   Emergency Actions: This describes actions that need to be taken immediately on the onset of the disaster, automatically, to protect life and property.

(9)   Transition: This describes how the company will transition from the immediate automatic actions to a more deliberate recovery process.

(10)   Recovery: This defines priorities for damage assessment and recovery actions, including determining which functions should be recovered first, what specific facilities, supplies, equipment, and communications will be needed, and the crisis communications strategies to be used.

(11)   Plan Maintenance: This defines the cycle for exercising, testing, review, and revision of the plan.

The focus of business continuity plans has often been restricted to issues of protecting the information technology infrastructure of a company. Although the problem is wider than just computer issues, the need to provide continuous data services has resulted in the development of a number of back-up facilities:[147]

Cold Site: A cold site is empty office space in which the business can establish a temporary data center.

Warm Site: A warm site provides a facility with operational computers and workspace, but company specific software must still be installed and tested before operations can commence.

Hot Site: A hot site is a facility with a complete operating environment similar to that of the business, including workspace and computers with the same operating and data management system loaded and ready for operations when a recovery team walks in with the most recent data in tape or disk form.

Mobile Site: A mobile site is a transportable facility (typically in a trailer configuration) with required hardware, software, and work stations, that can be moved to the disaster location to allow on-site restoration of critical business applications.

Electronic Vaulting: Electronic vaulting solutions provide for a continuous, near real-time transmission of key data from the business to an off-site location. The most robust version of this includes a complete staffed alternate data center ready to immediately assume operations in case of a failure of the primary site.

Pre-Plan

Pre-Plans are most commonly used by fire departments. A Pre-Plan is prepared for a specific facility as part of a routine process of surveying and planning for a response district prior to an emergency.[120,186] The Pre-Plan provides guidance to responders as to:

... hazards in a specific facility, including not only the contents but also those resulting from location, construction material, and building design,

... layout of the facility, including floor and vertical plan, accesses, fire connections, utilities, etc.,

... number of people working in the facility and significant life hazards,

... response routes, staging, and traffic flow, and

... standard response in terms of numbers and types of vehicles and people and where they will come from.

The format of Pre-Plans may vary by emergency department, but a typical fire Pre-Plan may include:

(1)   Data sheet: with detailed written information about the facility.

(2)   Site plan: an overall view of the site and its surrounding area.

(3)   Floor plan: for each floor of buildings.

(4)   Roof plan: for the building in case operations have to be conducted on the roof.[124]

Pre-Plans are typically developed for every major facility (including hotels, institutions, office buildings, retail establishments, restaurants, and industrial facilities) and for any facility that has the potential to pose unusual problems in a jurisdiction.[124] The completed document is typically one to four pages in length, uses a standard format, and makes maximum use of diagrams and tables of information to communicate key data quickly. Completed Pre-Plans are normally bound in a binder and stored in response vehicles or are available on computer systems (these can be produced and indexed simply using the capabilities of web page building and database softwares[60]). Some older systems of Pre-Plan design may still use an index card based system.[120]

Although Pre-Plans are typically used by fire departments, they can be used by any organization that can define a specific site to which response can reasonably be expected. Some examples include:

... an industrial plant emergency brigade pre-plans for emergencies in each building,
... a search and rescue team pre-plans for ramp searches for false emergency locator transmitter activations at a local airport,
... an emergency communications team pre-plans a hospital for response to provide radio communications in case of a telephone system failure,
... an emergency medical services agency pre-plans for a sports arena, etc.

Operations Plan/Order

The U. S. Armed Forces use a standard format for both Operations Plans and Operations Orders.[67,227,229,234,268] In this system a Plan provides general guidance and an Operations Order translates the Plan into specific instructions for a particular operation. Although the military Operations Plan format is not commonly used in the civilian world, the concept of an Operations Order should be considered. Many types of events demand considerable planning in order to have a successful outcome, and the use of an Operations Order provides a formal document as the specific, tailored plan of action for the event.

The Operations Plan/Order format uses five standard paragraphs with Annexes to provide specific details. The standard paragraphs are:[237]

(1)　Situation: The Situation paragraph describes what the current or expected situation is **B** who is involved, what is happening, where is it occurring, when is it going to occur, etc. The Situation also includes what resources are available from other organizations. And it provides the assumptions the planner has made about the situation and the response.

(2)　Mission: This is a short statement of what is to be done, when, and what the desired outcome is. Short means one or two sentences.

(3)　Execution: The execution section provides a general concept of operations. The mission and a well stated concept of operations communicates the course of action so that subordinates can achieve the desired objectives by applying good judgment even if the situation changes significantly. In addition, the Execution section assigns specific responsibilities to all of the participants and describes how their actions will be coordinated.

(4)　Logistics: The Logistics section addresses in detail all of the expected support requirements to sustain the emergency operation, assigns responsibilities for support, and identifies sources of support.

(5)　Command and Signal: For Signal substitute communications. This section identifies the chain of command and continuity of leadership for the operation, identifies the location of Command Post or Emergency

Operations Center, and identifies what communications systems and procedures will be used.

Another useful concept found in military operations is the Warning Order. A Warning Order is a much-shortened version of the Operations Plan/Order designed to give members of a unit as much information as possible as soon as possible to allow effective preparation for an assignment.[268] A Warning Order is a first look, often with incomplete information, at the task that will eventually be represented as an Operations Order.[15]

Incident Action Plan

An Emergency Operations Plan or Business Continuity Plan addresses, in general terms, how to respond to emergency events. When actually involved in the emergency, the Incident Commander must tailor guidance in the Emergency Operations Plan to the realities of a specific event and provide personnel and organizations guidance for the tasks to be completed during the next operational period. The Incident Action Plan format provides a way to do that. Incident Action Plans can be developed as outline documents using tactical worksheet whiteboards, a common use at fireground command posts. In this format the Incident Commander will typically use the whiteboard or tactical worksheet to brief key staff and to direct overall operations. The more formal version of this is the Incident Command System Incident Action Plan developed and issued as a published document at major wildfires.[166,181]

The objective of an Incident Action Plan is to establish and communicate priorities for the control of the incident during this operational period, establish measures of success, and to assign personnel and resources to duties to carry out the priorities. The Incident Command System forms used to develop the Incident Action Plan are the ICS 202, Incident Objectives, ICS 203, Organization Assignment List, and ICS 204, Division Assignment List.[166]

The critical portion of any incident action plan is the Control Objectives (ICS 202) or Tactical Priorities (also commonly used on the wide variety of tactical worksheet designs). These are a statement of what the Incident Commander believes should be accomplished during the operational period, in priority order. Objectives or priorities should be stated in clear terms and be measurable so that all participants can easily determine if they are making progress. The Incident Action Plan should also clearly identify any hazards that may endanger the disaster workers and provide guidance on how to avoid those hazards.[166,181]

Completed Incident Action Plans perform several important roles. They provide a written basis for briefings; they can be provided to all supervisory personnel to ensure everyone has the same guidance for action; they help document the overall response to a long-term incident; they may be used to determine who may have received certain types of environmental exposures; etc. And when the event is terminated, they are a permanent record documenting that management of the incident was orderly, thoughtful, and based on the best available information at the time.

Mutual Aid

Although often viewed as an agreement or an operational procedure, effective mutual aid is the result of a formal process and becomes a significant planning feature.

Mutual aid is the supply of resources from one jurisdiction or organization to another during an emergency according to pre-existing established procedures and agreements and based on specific request.[91] A version of this, known as automatic aid, provides resources automatically to a neighboring jurisdiction when certain conditions are met.[125] Mutual aid agreements should be formal written agreements[54] that define what resources may be requested, the procedures for making requests, what the obligations are for payment, liability protection, personnel benefits, and the length of such requests. Actual execution of mutual aid requests during emergencies depends on training and exercising and the use of standard terminology to identify:

(1)   The specific type and quantity of resources needed (requests on the order of "send everything you have" are difficult to source and may well end up with the wrong resources on the ground).

(2)   When the resources are needed.

(3)   The specific location to which they should be sent, along with a point of contact at that location.[210]

Although most mutual aid is local in scope, there are two significant exceptions. A number of states have developed statewide mutual aid agreements, either resource specific (as for wildland firefighting)[87,210,267] or as broad pacts for supply of all types of jurisdiction resources.[55,84] In addition, a state-to-state mutual aid agreement, the Emergency Management Assistance Compact, provides for states to supply a wide range of resources to any other state that is a signatory to the Compact.[6,51,212]

Standard Operating Procedure

A Standard Operating Procedure is a document describing in detail a standard way to perform a task that can be expected to be performed in any type of emergency.[59] The focus is on performance of routine tasks the same way every time they are done.[58] However, it is important to recognize that Standard Operating Procedures for commonplace incidents will be practiced to a higher level of familiarity than those for uncommon, and often larger, operational problems. This requires that training address when to implement Standard Operating Procedures designed for major emergencies and disasters.[271]

Standard Operating Procedures can be prepared in a variety of formats. Some organizations publish one detailed book that includes all Standard Operating Procedures. Other organizations publish procedures as individual documents, which are then inserted in a three ring binder. Because Standard Operating Procedures should be reviewed and revised on a relatively frequent basis, issuing them as individual documents has the advantages of making the revision and publishing process considerably simpler.

There is no commonly accepted standard format for a Standard Operating Procedure. However, an organization should adopt a standard format for its own documents so that users will find the same sorts of information in the same places in each procedure. As a minimum every standard operating procedure should include:[59]

(1)   An identification of the organization issuing the standard operating procedure.

(2)   A date of publication or revision.

(3)   A title.

(4)   A statement of scope and purpose that identifies for what the Standard Operating Procedure provides guidance and to whom it applies.

(5)   An explanation of the procedure, including, if appropriate, diagrams.

The actual text of the Standard Operating Procedure should be arranged in a format that makes it easy to follow during its use in an emergency and should be written in a to-the-point telegraphic style.   Generally, each procedure should be kept as brief as possible while still including all information a user would need **B** one or two page procedures are far better than five or ten page documents.

Typical Standard Operating Procedures used in a command center might include the following:

... incident log - when to start the log and how to fill out the individual blocks on the form (works well if each log sheet has this instruction page printed on the back)
... alerting - what methods to use, when alerts will be issued, how to document completion, what to do about individuals not contacted
... severe weather watch procedures - what to do when the National Weather Service issues a Watch for severe thunderstorms, floods, tornadoes, etc.

One form of Standard Operating Procedure that ties directly into the Emergency Operations Plan and site Pre-

Plans is the standard dispatch assignment,[204] also known in its earlier paper form as a run card.[72] The standard dispatch assignment assigns specific types and numbers of resources for each type of emergency incident, and based on location, assigns first-in units and subsequent increments of resources (known colloquially in the fire service as "second alarm," "third alarm," etc.). Standard dispatch assignments are now largely embedded in Computer Aided Dispatch systems.

Checklist

Plans define general guidance on how to respond to an event. Standard Operating Procedures tell how to perform commonly performed tasks. Checklists translate the guidance in Plans and Standard Operating Procedures into a step-by-step guide for action[263]. Checklists are designed to be used by individuals responding to an emergency to guide the completion of all routine tasks that must happen for the emergency to be controlled **B** usually for tasks that must be done rapidly under pressure. Checklists may be all hazard (for example, "Starting the Emergency Generator") or highly specific (for example, "Tornado Warning").

Checklists are typically short **B** even a complicated checklist would rarely be longer than two pages in length. The Checklist should be clearly identified with a simple title that communicates the type of actions covered and the circumstance for the use of the checklist. Actions that must be taken should be presented as short, bullet statements intended to jog the memory of someone already familiar with the task to be performed. The actions should be presented in the order in which they will be taken under optimum conditions.

Figure 5-1. Section of a Checklist

CHECKLIST 1 – EMERGENCY ACTIVATION:

_____ 1. Open incident log. Record all actions taken and information received.

_____ 2. Confirm incident details:

_____ a. Nature of hazard.

_____ b. Expected or actual time of impact.

_____ c. Assistance needed from our organization.

_____ 3. Alert initial duty shift.

_____ 4. Open the EOC (go to Checklist 5).

In preparing Checklists consider carefully the conditions under which they will be used. Checklists that will be used in an emergency operations center can be paper copies in document sleeves **B** those that will be used at a disaster site may survive better if they are laminated. Adjust the type font to make them easier to read; for example, 13 or 14 point type is much easier to read than 12 point. Critical steps can be emphasized through use of bold type or underlining or using a larger font.

The Checklist can become part of the key incident documentation. If Checklists are prepared with a lined section to the left of the actual items, times of start or completion of actions can be written in the blank. When using document sleeves, write on the sleeve with a water soluble audio-visual pen or (in field Command Posts) with a grease pencil. At the end of the event the completed Checklists can be copied on a photocopier to create a hard copy record before wiping the actual Checklist clean.

Other Formats

The Field Operations Guide has become a popular format for use in the emergency services. Some of these are essentially standard operating procedures books reduced to bullets and shrunk down to pocket size.[1,82,153,247] Others are closer to a checklist model with very abbreviated bullets intended as reminders of how to perform the procedure.[135]

The larger Field Operations Guides include a significant amount of reference material.[82,247] When early examples of Field Operations Guides, such as the 1983 ICS-420-1,[1] are compared to more recent documents, such as the Emergency Response to Terrorism Job Aid,[259] there is a clear growth from a simple checklist of responsibilities into a complete reference manual. Another approach to reference material is the quick reference card which provides few instructions, but does include as much detail as possible on the subject matter on a laminated pocket sized card.[225]

Using the Document

A single common failing in the management of crisis responses is failure to read or follow the Plan, Standard Operating Procedure, or Checklist.[18,59] Even when these documents are located in front of the staff in an Emergency

Operations Center, the average staff member will not open or use them. Planning and guidance documents contain a tremendous amount of information and considered thought on how best to deal with the effects of a given situation. They essentially take care of all the routine decision making so that the crisis managers can concentrate their thought on the unusual and difficult unique problems in a specific event.

Use of any of these documents is simple:

(1)   Each individual who will have a role in managing the emergency should be issued a copy of the Emergency Operations Plan and other appropriate planning documents.[81]

(2)   Locate the correct type of document for the situation. Does the situation require general guidance and reference materials for decision-making? If so, select the Plan. Or is it time to perform a routine procedure that applies to every event? If so, pull out the Standard Operating Procedures. Does the task require the individual move quickly with no time to think about needed actions? If so, grab the Checklist.

(3)   Open the book and find the right section. To do this in an emergency staff members must be familiar with the documents and have practiced with them in exercises. For example, a Checklist book may have several dozen Checklists – individuals should know the events for which there are checklists and how to find them in the book.

(4)   Read and follow the document.

(5)   Keep coming back to it to make certain all actions are completed and that the results are as expected.

(6)   Document completion in the incident log. Entries such as "EOC Activation Checklist completed at 1027" or "all communications watch phase preparations completed per the Emergency Operations Plan" convey that the Plan has been followed and help establish that actions were neither capricious nor negligent.

The Planning Process

Most guidance documents are prepared as a staff process well before an emergency event.   However, in extended emergencies a considerable amount of planning may be needed to effectively manage day-to-day operations.   As noted above, the resulting document is normally an Incident Action Plan.

Preparation of a Plan during on-going operations is a difficult process.   Only limited information on the disaster and its effects may be available.   Staff members will be focused on performing their emergency staff duties and may see time for planning as time they do not have.   The situation may be evolving rapidly enough that determining what to do in the next operational period is difficult.   Even determining what format to use may be difficult.

This means that effective planning during the event depends on three factors: (1) a standard planning cycle, (2) a simple format, and (3) the ability to distribute the work so that it can be completed rapidly.   The simple format means in practice the use of a standard form with blocks for completion of a pre-defined set of information.   This can be rapidly filled in by hand or on a template in a word processing or specialized emergency management software.   This standard format eliminates the need to think about what categories of information should be in the plan and helps ensure planners will not overlook critical information in the press of operations.

Actually completing the plan requires a regular cycle so that all participants know when their inputs are expected and when they can expect to see a completed document. Typically planning for the next operational cycle should start at the beginning of this one. Note that there are a variety of models of how this planning procedure will work – the steps listed below are a synthesis of a variety of approaches:[166,181,237,268]

(1)   Review the plans and situation reports for the previous and current operational periods to determine the flow of operations, what has been accomplished, and what still needs to be done.

(2)   Gather information on the current situation and obtain any forecasts of changes **B** pay special attention to any factors that will impact safety.

(3)   Determine what resources will be available during the next operational period.

(4)   Develop draft objectives, coordinate with appropriate section chiefs, and obtain executive leadership guidance and approval on overall objectives for the next period.

(5)   Hold a short planning meeting with key staff sections to brief what is known, communicate the overall objectives, assign responsibilities for individual plan sections to staff sections, and set deadlines.

(6)   As inputs come in from the staff, cross check them against other inputs to make sure the results are consistent and are in the same format, and then plug them into the correct section of the form.

(7)   When a draft is completed, circulate it to the other staff sections (actually walking it around is time consuming but demands attention and gets a better grade of feedback) for review.   If time and staff permits, consider tabletop exercising the plan to make certain it makes sense and is executable.

(8)   Complete the final form and obtain executive level approval.

(9)   Publish and distribute in the required number of copies (at least one copy for each staff and field unit supervisor, several for each staff section, several copies for file, etc.).

(10)   Brief the oncoming shift on the plan.

6.   STRATEGY AND TACTICS

Policy, Strategy, Tactics, and Task Level Decisions

Discussions of strategy and tactics are framed to some degree by the size of the organization. Large and small size organizations have to make the same types of decisions with the same level of relative impact on the organization. The primary differences in the decision process are the size of the problem, area covered, and number of resources to be considered.

Policy

Policy is a course of action on the part of an organization that is intended to produce certain expected results. Policy may be established by legislation, administrative regulations, court decisions, administrative decisions of officials, or the issuance of executive orders. Policy may be written, oral, or established by long custom and tradition. Generally, policy is seen as being relatively broad in scope, as applicable to whole classes of actions and problems, and as providing guidance for how an organization will carry out its fundamental responsibilities.[184,251]

Two levels of management are involved in emergency operations, policy setters, who determine what policy is or should be, and policy implementers, who carry out the established policy. Generally policy setters are executive level officials; in the public sector they are the senior elected (mayor, county commission, city council, governor) and appointed officials (city manager and county manager, state cabinet secretaries) of a jurisdiction. Policy implementers are all other employees.[251]

In most cases, crisis operations managers in most organizations are charged with recommending and implementing policy, not with formally setting policy. However, how policy is carried out operationally determines what the policy is. An organization may have a policy that stresses ethical treatment of the public; if, however, employees practice preferential treatment, discrimination, or other unethical practices, the policy becomes to others what the employees make it to be.

Strategy

Strategy in disaster response is guided by policy and is a conscious result of decisions about six key factors, normally guided by the Emergency Operations Plan:[184]

(1)   The nature of the threat.   All strategic decisions in emergency response must be based on an assessment of the size and complexity and the potential impact of the threat posed by a hazard.

(2)   The environment.   Terrain, weather, power, communications, roadways, and a variety of other environmental and infrastructure considerations limit or facilitate response.

(3)   Available resources.   Because strategy is about resource allocation, the decision makers must know what resources are available within the jurisdiction or organization, what resources are available by mutual aid or contract, and what resources are available from higher levels of government.   Knowledge of resources includes their capabilities and probable response times.   Again, this

knowledge is jurisdiction wide, and not restricted to the resources of one organization.

(4) Time. Although time is a resource, in emergency response it is of such importance as to require separate consideration. Time defines how much work is possible, what amount of work is required to control disaster effects, whether control is possible (for example, if it takes eight hours to build a levee and the river will crest in three hours, levee building is a social activity and not a viable strategy), and when resources are available.

(5) Supportability. Logistics concerns define whether a strategy can be carried out. For example, if the strategy includes opening shelters, but there is no way to feed the shelter occupants over the first 72 hours they may be in the shelter, a sheltering strategy may create more problems than solutions.

(6) Objectives. Objectives are defined by policy. For example, it is commonly accepted that it is reasonable policy for governments to attempt to protect life, property, the environment, and the economic and social fabric of jurisdictions. For corporations, the protection of shareholder value is a fundamental policy. Objectives logically flow from these policies. For a city, evacuation of residents downwind of a chemical plant accident is an objective driven by the policy of protecting life. Objectives at every level of response need to be framed in a way that is measurable.

Strategic decisions in a disaster are high-level decisions that translate policy and the guidance in plans into the actual presence of people and equipment on scene to deal with the disaster impacts. Strategy is a chief and senior executive level process of deciding how to allocate resources to achieve objectives in responding to multiple problems that threaten the community.[83]

For example, if our community historically is threatened by flooding, it is a policy decision not to build a levee because we wish to preserve our waterfront, but to attempt to protect life and property during an actual flood. Our possible strategies in this situation are two: build temporary levees of sandbags when needed, or evacuate.

In the actual event, the rapid rise of the river and the limited resources may mean that it is impossible to carry out both strategies. We simply may not be able to sandbag quickly enough or have the resources available to assist business in evacuating their equipment, records, stock, etc. However, we can protect the people. This means that the modified strategy becomes "to evacuate the residents of the Belle Meade section before the flood hits." To carry out this mission the strategy is translated into specific guidance: "the Police Department, City Emergency Medical Services, and the School Board, supported by public information announcements using the Emergency Alerting System, will evacuate the population of the Belle Meade section east of Highway 34 by 2:00 pm."

Tactics

Tactics translate the broad guidance of strategy into individual assignments of units and people to tasks. Tactical decision-making identifies:

(1)   The details.   Tactics look at solutions to problems in relatively small areas in which the fine detail

can be identified. A tactical decision may be as detailed as what software to load first in starting to recover company operations, or to which side of a building to deploy a rescue unit.

(2)   Specific resources. Tactical decisions involve specific resources (the Data Center Recovery Team, Engine Company 2, etc.).

(3)   The tasks. Tactical decisions assign resources to tasks - specific amounts and types of work that need to be done.

Tactical decisions are typically made at the Incident Commander, Operations Section Chief, Branch Chief, or Division or Group Supervisor level. Tactical decisions are incorporated in assignments in the Incident Action Plan.

To continue with the flood example, the Incident Commander for the Belle Meade evacuation might make a tactical decision that "all available emergency medical services units, with the exception of two ambulances, will evacuate the Shady Grove Convalescent Center — we will hold two advanced life support units in reserve to evacuate any home health care patients in the evacuation area that cannot be moved by their primary care givers." In this case we have assigned two specific tasks to the Medical Group and made a decision about how many of the available resources will be committed to each task. The Medical Group Supervisor makes a further tactical decision and allocates City Medic 7 and Transtar 21 as the reserve, ordering the two ambulances to take up posts to cover the area as the evacuation progresses.

It is important to note that the terms "strategy" and "tactics" are used differently in many emergency services organizations. For example, writers on fire suppression commonly talk about decisions as to whether to fight a fire from inside the building in an attack mode or to fight it defensively from outside the building as strategic decisions.[28] Two alternative uses of strategy and tactics, one of which is from English practice, are reflected in Table 6-1.

Table 6-1. Alternative Uses of Strategy and Tactics[83,154]

Term	United States Fire Departments	United Kingdom Fire and Police
Strategy	Incident Commanders establishing major objectives, priorities, resource allocations and determining mode of operations	GOLD Level – senior executives at the Headquarters Incident Room determining overall policy and meeting needs of tactical commanders
Control	Operations Officers supervising major segments of the organization	
Tactics	Sector Officers in charge of grouped activity in assigned areas or performing specific functions	SILVER Level – Incident Commander of a major incident located in the Command Post
Tasks	individual Company level activities	BRONZE Level – initial Incident Command and supervision of routine response

Note that the Bronze, Silver, and Gold English model is closer to the definition of strategy and tactics suggested for disaster applications. In the context of disasters, the quantity of resources involved, the number of agencies responding, the scope of the problem, and the level of threat to the community involved in most single incidents (within the larger disaster) do not elevate these problem to the level of strategic decision making. Tactical decisions cover a wide range of levels of decision-making and can address large problems with many response units, or small ones with only one unit.

Tasks

Tasks are actions taken at the level of individuals and small teams to actually perform work in an emergency situation. How to perform tasks is usually defined by Standard Operating Procedures and training for the specific task. The actual mechanics of performing the task for a given situation is determined by the individual worker or the immediate front line supervisor. While most of task performance has been established as a standard routine, how that task is performed should still be guided by the intent of the strategy and tactics for the incident and the threat that is being faced.

To use the flood example, a task is the actual process of going house-to-house delivering the evacuation notice. If the tactic is to evacuate the residents closest to the river, the task is to start at the low point on the street and work toward high ground. The Standard Operating Procedure for evacuations tells what to say to residents and what the content of the information flier should be.

Decision Processes

Determination of strategy, tactics, and tasks requires the emergency manager to make decisions. Some of these decisions are made in advance through the planning process; others must be made during the event itself.

The Decision Environment

The environment in which decisions are made often places significant constraints on the decision makers. In general, the following are desirable preconditions for effective decisions in major emergencies:[106]

(1)   Available time must be sufficient to allow a decision making process to be carried out, including the gathering of information, formulation of courses of action, and testing those courses of action to select the best one.

(2)   The information available must be of sufficient quality to allow good decisions.

(3)   The decision maker must have the intelligence, training, and experience needed to make the best decision for the conditions.

The issue of adequate information particularly deserves attention. All decisions based on information are made in one of three states, as shown in Table 6-2. Decision making under conditions of certainty is rare within the envelope during which most operational decisions are made. There are, of course, situations under which certainty exists, but they tend to occur when the outcome is obvious to all but the most perception challenged of decision makers. Most decisions are made in the range of risk or uncertainty.

Table 6-2. Decision Information Situations[191]

State	Probability	Outcome
Certainty	known	known
Risk	can be determined	can be predicted
Uncertainty	cannot be determined	cannot be predicted

Associated with this is a problem of information gathering. Theoretically, the more one learns about a situation, the greater the likelihood of reaching certainty in defining the outcome for a course of action. However, the cost of gathering the additional information to move from uncertainty to risk in terms of time and effort and lost opportunity may be high. With each increment of reduced risk, the cost will increase further, until it is no longer sensible to put off a decision.

Decision Processes

The Problem Solving Process

One approach to making decisions is to treat each decision situation as a problem to be solved. Although there are many descriptions of problem-solving methods, the following steps are common to most:[61,109,138]

(1) Gather information. The decision maker needs sufficient information to understand in broad terms how the event is evolving, what has happened to date, available resources, and actions in progress.

111

(2)   Identify the problem or the core issue. This is the critical point in the process; in many cases the decision maker may end up solving the wrong problem.

(3)   Develop possible courses of action that have reasonable chances of success.

(4)   Test the courses of action against known information and select the best course of action based on criteria set by public policy or embodied in the Emergency Operations Plan.

(5)   Implement the solution.

(6)   Monitor the results and adjust the course of action as necessary.

Research suggests that experienced Incident Commanders apply this analytical approach in no more than 10 percent of emergency situations.[106]   There are a number of possible explanations for this, including lack of time and resources.   However, for large, complex situations, the problem solving approach, even carried to the length of testing with tabletop exercises[268] or what-if discussions,[176] has a proven track record as the most effective way to bring the abilities of a staff to bear on the problem.   Although decision makers may make decisions in other ways, this does not invalidate the quality of a problem solving approach.

Intuitive or Naturalistic Decision-Making

Naturalistic decision-making by experienced decision-makers is based on matching a given situation that the decision-maker recognizes as typical or as similar to other situations with an appropriate course of action.[83] This process is known as recognition-primed decision-making. It depends upon four components:

(1)   Expectancies of what may occur.

(2)   Plausible goals that the decision-maker can expect to accomplish.

(3)   Relevant cues that place this decision in a familiar context.

(4)   Typical actions that the decision-maker's experience indicates have a reasonable probability of succeeding in these conditions.

In this process the decision-maker generates possible solutions serially and evaluates each mentally. There is an expectation that the first solution will be acceptable and that subsequent efforts will improve that decision.[106]

The Recognition-Primed Decision model suggests that there are three basic decision making situations, each with its own series of actions:

(1)   Level 1 – Simple Match: The decision maker identifies the situation as being a typical one and applies a response that he or she knows is the appropriate one.

(2)   Level 2 – Diagnose the Situation: The decision maker in this case does not immediately recognize the situation as being typical and has to consider several

interpretations of the situation, or even combine characteristics, to arrive at a good match.

(3) Level 3 – Evaluation: In cases of even greater uncertainty the decision maker may, after arriving at a course of action, mentally replay the course of action before implementation, and modify components if necessary.

This process is highly recommended as being of great value under conditions of uncertainty, with situations that fit the decision maker's experiences, and in which a rapid decision is required. The clear deficiency is that decision-making is concentrated in one individual, to the exclusion of use of the synergy of staff processes, and is largely dependent on his or her capabilities and experience.

Decision Aids

Decision aids provide guidance to decision makers in making routine decisions that have to be made in every type of event and in structuring unusual decisions. The most basic decision aids are Plans, Standard Operating Procedures, and Checklists. Plans provide policy guidance that should allow the crisis manager to make decisions consistent with the intent of the executive leadership of the organization. Standard Operating Procedures and Checklists provide step-by-step guidance on how to do tasks during an emergency so that no, or minimal, decision-making is required for most routine decisions. Some emergency management and business continuity softwares provide decision support, usually in the form of Checklists and Plan access.

The simplest decision aid is a positives-negatives chart. This is no more sophisticated than listing the advantages to a course of action in a column and comparing them to the disadvantages of the same course of action in a parallel column. This helps the decision maker visualize the impact of a decision and determine if the positives outweigh the negatives.

Table 6-3. A Simple Positive-Negative Comparison

Decision Question: The time is 10:30 pm. Do we order an evacuation tonight?

Advantage:	Disadvantage:
more time for evacuees to clear the area before storm impacts arrive	some people may not receive the evacuation message until the morning
better chance that everyone will get out	hazard of night time driving and potential for accidents
may get ahead of the traffic flow issues by getting some people on the road early	evacuees will be more tired with increased accident potential tomorrow
gives us a hedge in case the storm intensifies or speeds up its speed of advance	sheltering arrangements are not in place
	more of a danger to public safety staff
	may be hard to get the neighboring jurisdictions, highway department, state police mobilized

Another simple decision-making tool is a decision tree. These can be diagrammed quickly to capture the likely outcomes are of any decision. Understanding the impacts and the further decisions that will be needed helps determine the best set of decisions.[61]

Figure 6-1. A Decision Tree

Decision Question: Do we order an evacuation?

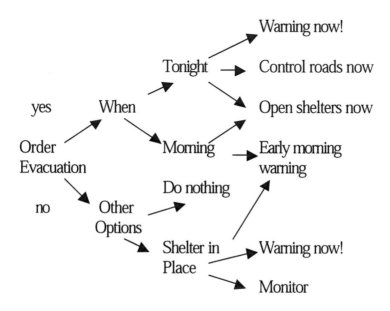

Other formats can be used to achieve the same objective as the decision tree. These include flow charts and varieties of flow charts used to depict algorithms in the emergency medical services.[14]

A third tool, the Time Delineating Schedule, offers a framework for organizing the phasing of decisions in a developing emergency.[202] This tool identifies typical decisions made in each phase, identifies the appropriate staff section to manage the decision process, and assigns a precedence. Originally designed for hurricane decision-making, it can be adapted to more rapid onset events and to any desired phasing and precedence structure. Table 6-4 shows a partial example of a Time Delineating Schedule for an Emergency Operations Center organized in a traditional model of groups, using the disaster phases suggested in Table 1-3 and the precedences for action in Table 4-2.

Table 6-4. Section of Time Delineating Schedule

Phase	Action Decision	Group	Precedence
Mobilization	Determine potential impact area	Disaster Analysis	Priority
Mobilization	Identify requirement for prepositioning	Operations	Priority
Mobilization	Select resources for prepositioning	Operations	Priority
Mobilization	Choose time for prepositioning	Operations Policy	Priority
Mobilization	Determine time to augment emergency operations center	Emergency Manager Policy	Routine

Population Protection

Warnings, Watches, and Alerting

In general, the first response step in protecting the public is in alerting the population that a potential emergency

117

exists that may pose a threat. Two terms are regularly used by the National Weather Service in alerting the public. Watch indicates that conditions are favorable for the development of severe weather and that a prudent course of action is to monitor weather conditions. Warning indicates that severe weather is happening at this time and immediate actions should be taken by persons in the impact area to protect themselves.[231-233] Similar terminology has not been developed for technological hazards.

A variety of methods exist to alert the public to emergency conditions. Sirens remain a viable technology and are still commonly used around nuclear power plants.[10,64] However, many community siren systems installed for nuclear attack warning have been removed. In high noise areas, especially in industrial settings, and in areas with large numbers of sealed buildings, sirens may be difficult to hear. Sirens do have areas of poor coverage and new residential developments may not be covered by sirens sited in the 1950s and 1960s. Other alerting methods include automated telephone alerting systems and the Emergency Alerting System, addressed in the chapter on Communications.

Protective Actions

A significant area for policy and strategy decision-making is the selection of appropriate measures for population protection during disasters. While this is normally discussed in terms of governmental actions, similar concerns should be addressed by business continuity planners in terms of the protection of the work force in their facilities and the protection of families of the staff.

Evacuation

One of the most common population protection measures is the ordering of an evacuation. Evacuations work by removing people from the impact area of the disaster and moving them to a safe area until the threat has passed.[65,117,125,184] Evacuations may be voluntary, usually issued as an advisory for people to evacuate well before an event is an immediate threat. In many jurisdictions, legal authority exists to order mandatory evacuation. However, it is impractical to devote resources to forcibly remove residents or business owners from their property. The mandatory evacuation order to some degree may absolve government from a responsibility to undertake risky rescues and shifts liability to the individual who does not evacuate.

An evacuation decision must be made early enough to permit evacuees to depart the impact area before the arrival of the impact of the event. This means that decisions must be made in advance of the amount of time that it will take to clear the area (called the clearance time). In the case of hurricanes, this decision time can be determined by placing a decision arc for the appropriate strength of storm and clearance time on a tracking chart. If the decision is made after the storm crosses this arc, risk of an unsuccessful evacuation increase. These decisions are complicated by the need to make decisions in time for normal news broadcasts, preferably by the 6:00 pm news.

Clearance times are typically based on the expected throughput of the available road system. These times must be adjusted regularly as the road net and the population changes. Experience indicates that clearance times are normally computed to the jurisdiction boundary — they do not represent the time it will actually take evacuees to reach shelter, especially if multiple jurisdictions are being evacuated and if the population of one jurisdiction has to pass through another that is also evacuating. This increases the potential for large numbers of evacuees having to ride out the event's impact in their vehicles on the highway.

Vertical Evacuation

In some flooding scenarios, it may be impossible to evacuate all of the residents of a densely populated urban area prior to rising water eventually making evacuation routes impassable. In areas that may face this threat, vertical evacuation may offer a viable strategy for population protection. In vertical evacuation, people are sheltered above the water level in the upper floors of high-rise buildings. This presumes that buildings selected are well constructed enough to resist the effects of the flooding — for example, it is probably not a viable strategy in ocean front areas where buildings will be subjected to wave action or in areas where wind velocity may threaten integrity of the building envelope.[94,171]

Sheltering in Place

In many situations it may be preferable to direct the population to remain in their homes, schools, or places of work until a threat has passed.[3] For rapidly developing threats, sheltering in place may avoid exposing the population to the danger of being caught on the road. Sheltering in interior hallways or bathrooms may offer some protection; a greater degree of protection may be afforded by underground storm shelters or specially designed safe rooms.

For chemical or radiological releases sheltering in place includes closing windows and shutting down air handling systems. It is important to note that some populations may be better protected if they remain in place even when a long lead time is available — some home health care patients may fit this criteria if they can be provided with generators and if their homes are in reasonably safe locations.

Sheltering in Shelters

Public shelters offer evacuated populations (both voluntary and mandatory), and individuals who have had to leave their homes as a result of disaster damage, basic protection from the event and basic services (food, water, a place to sleep, and sanitation facilities).[184] Opening shelters represents a significant resource commitment in terms of shelter staff (whether this is provided by government employees or voluntary agency personnel), support and logistical services, and administrative services in the tracking of shelter residents. Shelters may be opened either as a result of threats to the jurisdiction or facility or as a result of a decision by another jurisdiction to evacuate (to accommodate individuals traveling through the jurisdiction).

To protect evacuees shelters must meet certain standards. These include the ability to resist expected winds, location out of flooding areas, and the ability to set up decontamination facilities (for nuclear power plant evacuation shelters). Because shelters may accommodate people for an extended period of time, shelters must have adequate sanitary systems. Activation of a shelter requires that some provisions be made to allow shelter occupants to sleep and that food and water be supplied to the shelter. A related issue is that sheltering animal populations, requiring provision for pet sheltering in proximity to human shelters.

Experience indicates that relatively few evacuees actually use shelters. Most either shelter with family or friends or use hotels or motels along the routes of evacuation.[125]

A special class of shelters are special needs shelters designed for people whose care needs exceed the level of service available in normal shelters.[65,184] Most persons with disabilities can be accommodated safely and relatively comfortably in normal population shelters.[255] However, for individuals with severe disabilities or who require constant medical attention and life support systems, it may be necessary to establish shelters in facilities that can provide the needed services and which can be staffed with appropriately trained care givers.

Refuges of Last Resort

Not all evacuees will be able to complete an evacuation or reach a safe shelter prior to the onset of severe impacts from the event. Some may be trapped on congested roadways or may have made a decision to evacuate from a hazardous location too late. In some cases, public safety or facility security personnel must remain in the impact area and ride out the event. The ability to protect any of these people will be very limited. However, some level of protection may be provided by refuges of last resort. These facilities may not offer the protection that a normal shelter offers, but still offer enough protection from the primary impact of the disaster that survival is possible with good luck and under primitive conditions. Identification of refuges of last resort along evacuation routes and for persons expected to ride out the impact may be a prudent strategy.[16,26,177]

Restoration of Operations

Continuity of Operations and Continuity of Government

In major disasters, there are two significant concerns that must be addressed, the maintenance of continuity of operations and continuity of government. Continuity of operations refers to the ability of the community and its vital services to continue to function. Continuity of operations requires the ability of the public safety services, basic government services, utility lifelines, and the economic and social infrastructure of the community to continue to function and to recover from the impacts as quickly as possible. Continuity of operations planning must include the development of priorities for the restoration of services, the pre-positioning and protection in place of emergency recovery resources, and the development of robust communications and utility restoration capabilities.[241,260] Continuity of operations planning should also include assurance to key workers that their families will be taken care of through evacuation or sheltering.

Continuity of government assures that the three basic branches of government in any jurisdiction, executive, legislative, and judicial, can continue to function after a major disaster. Continuity of government programs have focused on seven components:[252]

(1) Succession. Continuity of leadership is a particularly critical issue (for both government and

123

business), with procedures in place for the orderly transition of leadership to a rank order list of deputies in case the chief executive is killed, incapacitated, or unable to communicate.

(2)   Pre-delegation of emergency authorities to allow government to continue under emergency conditions.

(3)   Emergency action steps that provide guidance on the actions needed to control the emergency.

(4)   Establishment of an Emergency Operations Center.

(5)   Establishment of an Alternate Emergency Operations Center.

(6)   Safeguarding of essential records through their identification and protection so that they can be made available when needed.

(7)   Protection of government personnel, resources, and facilities so they will be available for emergency response.

Supply Chain Issues

The supply chain is critical to the rapid restoration of normal operations in a disaster. Few organizations stockpile sufficient supplies to allow continued operation when delivery of supplies are interrupted. For example, even state road departments may run out of road treatment chemicals in a major snowstorm. Most hospitals operate on just-in-time delivery of pharmaceuticals and soft goods and rarely stock supplies for extended periods. Priority must be given to identifying critical supplies, stockpiling what is feasible, and having contracts in place for rapid resupply during the response and emergency recovery phases. With this comes a priority for the rapid reopening of roads, airports, and other transportation facilities as part of maintaining continuity of operations.

7. EMERGENCY COMMUNICATIONS TECHNOLOGIES AND PROCEDURES

Communications Organization and Staffing for the Emergency Operations Center

Many jurisdictions operate a 24 hour Public Safety Answering point, or Dispatch Center. This facility provides public access to emergency services and communications links for the control of public safety response. This facility is not the Emergency Operations Center, although the Emergency Operations Center may be collocated and share facilities with the Dispatch Center.[209] Similarly, the communications capabilities of the Public Safety Answering point are not designed in most cases to meet all Emergency Operations Center communications needs.

Emergency Operations Centers may have their own communications capabilities and a separate Communications Unit to manage the variety of radio and other capabilities. The traditional model of an Emergency Operations Center Communications Center envisions two key functions:

(1)   A Radio Room with operators responsible for radio communications within the jurisdiction and long range communications to other jurisdictions and state agencies.

(2)   A Message Center with staff to process incoming messages, record them in a message log, send them to the correct staff section by runner, and to collect outgoing messages and process them in a similar way.

Although new means of communications augment the traditional radio systems, this same organizational model retains significant utility. Adding to the Radio Room a fax machine, a satellite telephone, and an e-mail capability upgrades it to a more modern standard. The Message Center and Radio Room still can perform the functions of (1) providing the primary means of record communications to other organizations (as opposed to staff coordination telephone calls) and (2) providing the primary backup communications circuits in the event of telephone failure.

Staffing this facility requires individuals familiar with radio procedures and with a wide variety of communications systems that require operator control. While Dispatch Center staff members have communications experience, they must be trained in radio message procedures and in the operation of the Emergency Operations Center's equipment.[195] However, requirements for augmented dispatch staffing during a major event[204] may make assignment of dispatchers to the Emergency Operations Center Communications Center function difficult. In some cases the staffing requirement has been met by recruiting a volunteer communications auxiliary for the Emergency Operations Center and training these individuals to manage the full range of communications systems.[31,170]

Effects of Emergencies on Communications

Role of Communications in Disaster Response

In disasters, communications systems perform the same functions required for day-to-day operations of public safety agencies. These include: receiving reports of emergency situations, answering assistance requests from the public, determining resource status, dispatching resources, coordinating and directing operations, obtaining situation assessments, coordinating with other jurisdictions and agencies, requesting mutual aid, reporting situation information to senior leadership and state agencies, and maintaining accountability and safety of resources. Figure 7-1 suggests some of the linkages required to meet these roles.

Figure 7-1. Selected Typical Local Jurisdiction Emergency Operations Center Communications Linkages

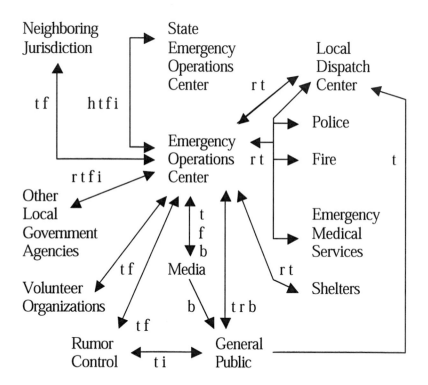

Note: r – radio including government, amateur and Citizens Band; t – telephone including cellular; f – facsimile; h – high frequency radio; i – Internet including e-mail; b – broadcast including the Emergency Alerting System

Physical Effects

On the most basic level, emergencies can seriously damage the equipment and facilities needed for communications.[218] High winds, flying debris, and ice can physically damage antenna elements or even cause antenna towers to collapse (for example, Hurricane Andrew destroyed six of seven backup communications antennas at the Dade County Emergency Operations Center[122]). Even if antennas remain in place, directional ones may be knocked out of alignment and require resiting. Heavy snow or even rain may interrupt signals from satellite dish systems. Lightning strikes, even on well-grounded systems with surge arresters, can destroy antenna elements, the feed lines that connect the antenna with the transceiver, or the transceiver (a combination of transmitter and receiver that makes up modern radio systems) itself. Flooding can fill underground Communications Centers and power systems with water; high winds can pick up gravel from roofs upwind and shatter above ground Communications Center windows leading to building failure. Power lines are vulnerable to wind, ice, and debris, and buried telephone lines may have vaults that flood, disrupting telephone service.

Load Issues

Disaster operations can create tremendous additional loads on communications systems[218] because of increased demands for service from the public, increased coordination with other agencies, and reduced system capacity due to damage.[190,210] The distribution of this load may be highly variable, in terms of services and communications circuits impacted, in the timing of the added load, and in the facilities that are affected. Factors that contribute to this variability

129

include the intensity, duration, and breadth of the event=s impact, the area impacted, and the time of impact. Very high impact events may actually reduce the volume of communications as a result of damage to systems. Similarly, late night impacts may generate relatively little communications load, especially at the state level, due to uncertainty about the extent or degree of damage.

Human Factors

Intensive communications in a disaster impose significant loads on the staff of an Emergency Operations Center. Several factors contribute to this. First, Emergency Operations Center staffs are drawn from many agencies and may be unused to the volume of calls and the variety of types of communications systems with which they have to work. Second, multiple phone calls and conversations make for a very noisy working environment, with a resulting increase in stress. Third, many of the individuals communicating are under high degrees of stress themselves, resulting in voices that transfer stress to the recipient of the call, in desperate pleas for assistance that cannot be met, or even in verbal abuse when service is delayed or denied.[44]

Planning for Emergency Communications

Effective communications planning to maintain the ability to operate through and after the impact of a disaster means hardening systems, having redundancy with many alternate ways of transmitting information, stockpiling backup equipment that can be installed as an expedient to replace damaged components, and developing backup power systems including generators and batteries. If possible additional capacity in communications circuits should be available on call to handle increased loads. Personnel who will have communications roles should be trained in standard communications procedures to reduce the learning curve during the event and to reduce the potential for confusion or for lost information. Use of standard terminology is particularly important in reducing confusion.[24] In addition, individuals who will staff an Emergency Operations Center should be trained in stress recognition and techniques to reduce its impact.

The increased volume of communications required in any major event requires aggressive management to reduce communications system overload and failure. One approach is to designate radio frequencies (or channels) for specific tactical uses on scene.[24] Units and agencies responding to an incident remain on the dispatch frequency until arrival on the scene. Once on-scene, they switch to a tactical frequency for the incident. As the incident grows in complexity, additional frequencies are allocated to either specific functions,[218] to geographical areas of the incident, or for command or support functions. Communications can be shifted to cellular telephones or other communications systems for some functions that do not require the multiple-user, immediate contact capability of the radio. The original on-scene frequency remains an operational channel, often for the most heavily committed units.[24]

Coordination of various units responding to a major event remains a significant issue, especially as the variety of communications systems increases. It is possible to have four agencies on scene with four communications systems, including low band VHF, high band VHF, 400 MHz, and 800 MHz trunked systems. In this scenario none of the four agencies can talk directly to any of the other three.[218]

131

Inability of responding agencies to communicate with each other remains one of the most common deficiencies noted in major emergencies.[90] The use of common regional and statewide emergency frequencies allows at least a basic level of communications compatibility between responding resources.

Communications planning should also consider how communications flow fits with the organizational structure in use. One of the principles of an Incident Command System is that the combination of resources into strike teams, task forces, divisions, and groups reduces communications requirements by reducing the number of communications users. The hierarchical structure of the Incident Command System reduces communications use by encouraging strict adherence to vertical communications with the chain of command[24] and discouraging horizontal communications.[218]

Communications Technology

Radio Systems

Radios work by transmitting waves of electronic energy through the atmosphere. The characteristics of these waves determine the radio frequency (based on the number of electronic cycles per second the radio generates) and the physical length of the signal (which determines the length of antennas). The signal generated by a radio travels in three ways. There is a direct wave, commonly called line of sight transmission. If it is possible to draw a direct line between the antenna of the transmitter and the antenna of the receiver the result is line of sight.

A second type of propagation of signals is by ground wave; a portion of the electronic energy actually follows the contour of the ground, allowing transmission over hills and over the horizon. Lower frequencies may have a significant ground wave transmission; those above the lower band of the Very High Frequency range generally do not. A third type of propagation is the skywave; a portion of the signal is reflected by the ionosphere and travels great distances. Although sky wave can happen in the lower half of the Very High Frequency range, it is the basis for most High Frequency communications. In general, the higher the frequency, the more dependent it is on the direct wave and the greater impact obstacles will have in blocking transmissions.

Frequencies in three frequency bands are commonly used in public safety communications:

1-30 MHz	High Frequency (HF)
30-300 MHz	Very High Frequency (VHF)
300-3000 MHz	Ultra High Frequency (UHF)

Public safety communications are conducted in the low (30-50 MHz) and middle (130-160 MHz) portions of the VHF frequency band (commonly called "low band" and "high band"), and in the 400 and 800 MHz ranges in the UHF band. Most very high frequency and all ultra high frequency radio systems use frequency modulation (FM) as the way to transmit information in the radio signal.

Radio systems in general use for emergency communications generally can be classified as falling into two different types, line of sight and long distance. Long distance communications used to depend solely on high frequency radios which are able to transmit their signals by bouncing the radio waves off the ionosphere. High frequency, when combined with near vertical incident skywave (an antenna design that directs radio energy directly upward to allow short range coverage), can provide excellent statewide radio communications. However, the useable frequencies vary throughout the day, and communications may be disrupted by solar activity. Generally, the lower the frequency, the greater the length of time it can be used during the day.

However, today wide area coverage can be achieved with line of sight systems through the use of linked repeaters, which allow a message to be transmitted to one repeater and to be retransmitted by that repeater to all others in the system. Similarly, packet data systems can transmit data messages over long distances through the use of a series of nodes and digipeaters.

Line of sight communications have long been used for tactical communications (those between individual response units on the same incident). These depend on direct wave transmission and are therefore limited by the height of the antenna (the range of the direct wave of a radio can be estimated very approximately by multiplying the square root of the antenna height by 1.2). Such simplex operations (sometimes called Atalk around@) depend on the transmitter and the receiver operating on the same frequency. As long as operations are simplex, the only way to extend range is to use relay stations, preferably on high ground or with tall antennas.

However, line of sight systems can be extended to cover large areas through the use of repeaters. Most repeater operations are half duplex. The transmitter sends its signal on one frequency and the repeater retransmits that signal on a different frequency for the receiver. A repeater located on a tower allows a station at he limit of coverage to send a signal to a station at the opposite limit of coverage - with a well-sited repeater the entire response area of an agency can be covered.

Telephone

The primary means of communications in most emergencies remains the telephone. Maximum flexibility can be achieved with multiple line units routed through an internal exchange. Typically these allow calls to be held, parked for pickup by others, sent to voice mail, and answered on a speaker-phone. Speaker-phones are particularly valuable in allowing multiple participants to listen to a briefing or a conference call. However, speaker systems should also be equipped with a mute capability to ensure that unintended background noise is not transmitted to other participants.

If an internal exchange system is used in a building or agency, it is vital to have at least one telephone line wired directly to the normal external exchange. Internal exchanges and multiple line sets depend on commercial electric power to power their functions. If power fails, a directly wired, standard telephone may continue to operate.[190]

Another type of telephone service is provided by lines that are configured to ring only at one location. Variously called "hard wired," "ring down circuits," or "hot lines," these connect one facility directly with another, and ring when the individual picks up the handset.[190]

Telephone Alerting Systems

Telephones can be configured for use as community alerting systems – one option is through the use of software at the emergency operations or communications center.[69,79,221] Another option is the use of an outside call center that will use its system to call all telephone customers within specified areas.[2] By simultaneously dialing multiple lines, these systems deliver an identical scripted message to all residents of an area that has been identified, typically based on standard maps,[69] as being threatened. Some products allow remote activation of the system.[221] The message allows more information to be transmitted than is possible through a television crawler or a siren.

Even with the use of large numbers of lines (systems are currently being marketed that will support 72[69] or 96[79] lines), this can be a slow process and requires precise choices of message and alerting area. Also, these systems require that someone be at the called telephone number in order for the warning to be transmitted. Finally, systems based on a central call center require regular practice to ensure passwords and messages are current.[2] However, such automated alerting can be a useful part of an overall warning system.

Fax

Facsimile machines provide the capability to transmit written instructions or reports from one location to another as a finished copy, without the potential for garbled understandings of a text read over a radio or telephone line. If a fax is being used for emergency notifications or the dissemination of situation reports, it should be a multi-line system that can simultaneously send copies to multiple phone numbers. The speed with which faxes can be sent is directly related to the number of pages of material. Therefore, material to be faxed should be restricted to one page, and cover sheets should not be used if at all possible. Two sheets effectively doubles transmission time. Computers with internal fax capabilities offer a tremendous efficiency by allowing you to fax directly from your word processor printer.

In an age of e-mail, the fax still has significant advantages in emergency operations. An e-mail emergency message can arrive after an individual has checked his or her e-mail for the last time during the day, or while the individual is out of the office. However, a fax message to most offices will at least gain the attention of administrative staff, and increase the chances the message will get to the decision maker. And in an environment where a percentage of Emergency Operations Centers lack computers,[100] the fax may be the most effective way of transmitting authoritative text materials. As with any means of communications there are potential failure points (as simple as the fax machine paper tray being empty), and receipt of critical fax messages should be confirmed.[190]

Cellular Telephone

Cellular telephone systems have become an accepted way for emergency services to communicate. However, there are certain system vulnerabilities that make it unwise to rely on cellular telephones as the only means of backup communications. Cellular conversations can be monitored using modified scanners. Cellular sites are often overloaded during emergencies by requests for service from cellular telephones, making it impossible to access them when needed.[190] As fixed infrastructure, cellular sites are vulnerable to disaster effects (although portable cellular sites can be deployed within days to restore or expand communications[5]). And, because a cellular system depends on the normal telephone land-line system to complete calls, a failure in the land-line structure means that the cellular system is ineffective.

Satellite telephones operate on the same basis as cellular telephones; access is to a satellite transponder instead of a cellular site. Contact can be either mobile to mobile using satellite and a ground station or mobile to satellite to ground station to a conventional telephone system number. Coverage is nearly universal - as long as it is possible to achieve a look angle to see a satellite, connection with the system results.[107] Because of cost of both equipment and air time, satellite systems do not suffer from the congestion of normal cellular systems.

Wireless Data

Wireless data systems allow laptop computers equipped with a communications card to access data communications networks without the necessity for direct connection to a telephone circuit. In public safety uses this is most commonly found as mobile data terminals in emergency vehicles.[218] However, wireless data can also support laptop computer operations including the use of electronic mail and faxing from emergency staff in temporary facilities or from an Incident Command Post.

Pager systems are another form of wireless communications.[80] In addition to their traditional use for alerting staff, pager systems offer increasingly sophisticated operational capabilities. Specialized information services, most notably tailored weather alerting, are now available by pager.[134] Pagers have become two-way communications tools with the introduction of capabilities for two way paging and e-mail transmission and receipt.

Like cellular systems, wireless data systems are vulnerable to damage to towers or to interruptions in power or connectivity. In addition, coverage is generally more limited than that provided by cellular telephone systems, with generally good coverage in major metropolitan areas and their suburbs areas, but poor coverage in rural areas.

Automated Vehicle Location Systems

Automated vehicle location systems match wireless communications with Global Positioning Satellite system data to continuously report vehicle locations to a Public Safety Answering Point. When combined with computer aided dispatch, this allows mapping of the vehicle and association of incident data with vehicle in near real time. As a result Dispatch Centers can track resources more effectively and select the closest resources for a specific incident based on current map data.[194]

General Public Radio Systems

Traveler Information Stations

The Federal Communications Commission allows government entities to operate low power radio stations to broadcast information intended for motorists in a jurisdiction. Whether located in a transportation department or in the Emergency Operations Center, these stations can be used to provide evacuation instructions, and a variety of other emergency information, to the motoring public.[140]

Emergency Alerting System

The Emergency Alerting System is a national system, including AM and FM broadcast, television, and cable television stations, for the purpose of disseminating emergency messages to the public. State and local government and the National Weather Service can use the Emergency Alerting System to broadcast emergency warnings and instructions by accessing their State Relay Source (state government) or Local Primary Source (local government) stations using procedures in the local emergency communications plan.[244]

Internet, E-Mail, and Chat

We are just starting to understand the capabilities and limitations of computer based communications in disasters. The use of electronic mail with predetermined address lists allows nearly instantaneous distribution of emergency messages, reports, and situation information. Internet pages can be established and easily updated to make current situation information available on demand to large numbers of people. Listservs allow rapid exchange of e-mail messages between all listserv subscribers. Chat rooms, if properly managed with a moderator, allow virtual staff meeting discussions between widely separated participants. All of these means create a potential archive trail if paper copies are printed for the incident record. And the development of wireless communications options for laptop computers increases the potential for their use as tools for communications and management of actual emergency response[211].

Two cautions must be understood, however. Internet sites are only as good as the individuals or organizations which post them. There are emergency related sites that contain second or third hand information or information that is simply incorrect or fanciful. Prior to using Internet sources you must evaluate the value of the data. One of the areas in which access to highly reliable data is proving especially important is the rapid availability of data on developing weather events.[216]

Second, it is important to understand that individual Internet sites are vulnerable to failure. System loading and equipment failure can slow or block transmission of e-mail; listservs can exceed system traffic limits (if set too low) and shut down; and huge numbers of attempts to connect can jam individual sites. Relying on Internet access to high visibility sites as the only source of emergency information is a risky strategy.

Television

Although a receive-only system, television provides significant incident information for the operations, analysis, and planning staffs. News media often broadcast from an incident scene early in the incident, and these broadcasts can provide significant additional, near real time, situation information. Televisions selected should have screens large enough to allow easy viewing given the configuration of the Emergency Operations Center.[209] At least one television should be configured with a Video Cassette Recorder for taping of news reports and disaster footage.[147]

141

Federal Government Communications Systems

The Government Emergency Telecommunications System (GETS)

The Government Emergency Telecommunications System (GETS) provides priority telephone access for GETS subscribers in an emergency. By dialing a specific telephone number and using an access code number, the GETS subscriber:s call is processed to provide a high probability of completion through the use of identification as an emergency call, the use of special trunking techniques, and exemption from network management controls. Federal, state, and local agencies with national security and disaster response missions may participate in GETS.[262]

Shared Resources (SHARES)

Federal agencies with high frequency radio capability participate in Shared Resources (SHARES), a network of high frequency radio stations that provides a way to pass federal agency emergency radio messages in a disaster. SHARES participants include the National Guard and the state emergency operations centers.

Amateur and Citizens Radio

Amateur Radio Technology

Amateur radio communications most commonly include four capabilities: line of sight voice communications, long distance voice communications, packet data transmission, and television transmission. Line of sight voice communications are provided primarily by the use of frequency modulated (FM) radios in the 6 meter (50 MHz), 2 meter (144-148 MHz), 1.35 meter (220 Mhz), and 70 cm (440 Mhz) bands. Of these, two meter is the most commonly used. Communications are typically either simplex (the radio transmits and receives on the same frequency), limited to nearly line of sight contact with the station called, or half duplex (station transmits on one frequency and receives on another). Half duplex operations require the use of a repeater, typically located on a tower, building roof, or mountain top to provide the greatest possible range.

Long distance voice communications are normally provided through the use of single sideband high frequency radios operating on assigned amateur radio frequencies in the spectrum from 1.8 MHz to 30 MHz. Voice transmission is the most common capability, although high frequency will also support a variety of data transmission protocols. Amateur operators licensed to operate in these frequencies must also have demonstrated the ability to use Morse code, although code operations are generally considered obsolete for emergency use.

Packet systems are most commonly based on line of sight VHF FM radios, with a series of nodes or digipeaters (a digital repeater) allowing data to be transmitted from computer terminal at the transmitting end to a computer terminal at the receiving end. Specialized modems, known as terminal node controllers (TNC), allow data to be stored or directed to other sites and accept a variety of different communications standards. The advantages of packet include the ability to send relatively long text messages with a high degree of reliability and the relative security of the system.

A related capability is the Automatic Position Reporting System, which uses data from a Global Positioning System receiver and packet technology to automatically report and map the position of an amateur station and track mobile stations.

Amateur radio also provides the capability to transmit television images to support on-going damage assessment or real time monitoring of events. Current technology allows these images to be downloaded to computers for inclusion in situation reports and incident records.

Amateur Radio Organizations

Three Amateur Radio organizations provide emergency communications capabilities that can support government crisis operations. Two of these are specifically emergency oriented and based on city and county organizations. The other provides long distance communications.

The Amateur Radio Emergency Service (ARES) is a volunteer membership organization sponsored by the American Radio Relay League, the largest amateur radio society in the United States.[115] ARES members provide a variety of public service communications support activities and can support governmental or voluntary organization needs for emergency communications on request. ARES generally provides local area tactical communications.

The Radio Amateur Civil Emergency Service (RACES) is a government sponsored organization, associated with local and state emergency management agencies, which uses amateur radio communications for disaster response.[220] RACES is activated by government specifically to support governmental requirements.[115] RACES capabilities include local tactical communications and linkages with the state emergency operations center.

A third amateur radio system, the National Traffic System (NTS), provides a capability to send radio messages over long distance using a combination of local, regional, and national traffic nets.[115] A message passed through this system can be delivered across the country in a day under ideal conditions. Much of the welfare inquiry radio traffic is handled by the National Traffic System.

An additional amateur radio capability exists in the Armed Forces through the Military Affiliate Radio System (MARS) which serves the Army, Navy, Marine Corps, and Air Force.[235] MARS provides a practiced communications system of trained operators used to handling formal written messages. Depending on service policies, this system may be able to handle a significant volume of traffic in a disaster. Major voluntary agencies may also have their own amateur communications operators - for example, the Salvation Army sponsors the Salvation Army Team Emergency Radio Network (SATERN).[200]

Citizens Band

Citizens Band, an unlicensed communications service in the 26-27 MHz range, provides some capability to interface with the general public and to receive calls for assistance on Channel 9, a designated national emergency channel. However, the number of Citizens Band users is relatively small, misuse of the emergency channel a frequent problem, and the lack of organized emergency operators a

limitation.   The sole remaining national level Citizens Band
emergency   communications   organization   is   REACT
International,   and   the   coverage   REACT   provides   is
limited.[192]

Amateur Radio and Business

Although ARES or RACES may support emergency
response actions to preserve the assets or assist the people of
a business, it is important to understand that amateur radio
cannot be used for business purposes.   Where this line is
drawn in an emergency is not well defined.   It is probably
acceptable for emergency management to station a RACES
operator to coordinate between a business crisis operations
center and the jurisdiction emergency operations center.   On
the other hand it is probably not acceptable for that operator
to pass messages to other offices of the business to arrange
for processing of orders.

Communications Formats

Voice Formatted Messages

One   option   for   transmitting   information   or
instructions, that will be the same in virtually every situation,
is the use of a voice format message.   These messages have
standard fields of information preprinted on the message
blank with a line for the applicable text for the specific
situation to be inserted.   The form organizes the information,
guides the communications operator to pass the information
in the same way each time, and helps the message originator
ensure that no critical information is omitted.[40]   When
transmitting the format the operator reads a line number and
the applicable text; the individual receiving the message

copies down the text on the numbered line of the same form and uses the preprinted information to interpret the final message.

Voice formats work well for situation reports, damage assessment, patient tracking, changes in alert status, and mission tasking orders. Although designed for voice transmission by telephone or radio, the form, when completed, can easily be sent by fax.

It is important to note that Standard Operating Procedures may specify a standard format for certain types of emergency radio communications, without the creation of an actual format sheet. Size-up messages transmitted by the first arriving fire company at a fire scene (unit identification, location, type of building, assessment of the fire) are an example,[24] as are maritime distress messages (Mayday, call sign, position, type distress, assistance needed, vessel description, persons on board) [243] and the call sign-location-injured-problem format suggested for Citizens Band calls for assistance on channel 9.[192]

Radio Messages

Formal radio messages, written down on a message form, and transmitted verbatim from one station to another, are commonly used in disasters to provide record communication for requests for assistance, directions to move resources, casualty information, and instructions to order or purchase supplies. A written radio message has the same general legal status as a telegram or a letter for documenting actions.

147

Formats for written messages vary widely. However, some general rules are applicable to any radio message:

(1)   Write short messages. The shorter the message, the less likely that it will be garbled in transmission or during copying.

(2)   Write telegraphically. Cut out all words that do not contribute to meaning.

(3)   Make certain full address information is provided. The station that receives the message may have to deliver it by telephone (include the telephone number) or by courier (include a physical address). Even though you may know exactly where it needs to go, the individual on the other end of the radio may not have the same understanding.

(4)   Sign the message. The signature documents the writer and provides authority for the actions directed. Unsigned messages may expose the individual radio operator to liability for content.

(5)   Assign a realistic precedence. Precedences determine the order in which messages will be transmitted by communications operators. The amateur radio precedence system is one of the most commonly used: EMERGENCY - for disaster related messages that involve safety of life and property or the movement of response units and that must be delivered as rapidly as possible; PRIORITY - for other messages about the disaster that can accept some delay in their delivery; WELFARE - for inquiries about the health and safety of individuals in the disaster area; and ROUTINE - for all other messages. Emergency messages are transmitted before Priority ones; Welfare messages are often

not accepted until several days have passed after the disaster impact.[115]

Radio Logs

Most modern public safety radio systems operate with some form of logging. In systems based on a central Dispatch Center, logging is normally done by a recording system which records radio transmissions and telephone conversations with either an automatic time stamping or by procedurally requiring dispatchers to state the time of the transmission as part of the communication. These systems may be tape or computer based.[189] However, in a disaster, communications may be conducted on a variety of frequencies which are not recorded, and recording systems may not be operating.

Like the radio message form, the radio log provides a written, legal record of a station's radio communications. Log entries should include:

    … date and time of transmission
    … station contacted
    … frequency or channel in use
    … a brief description of the substance of the communication
    … operator on duty.

While it is impractical for individual radio users or a vehicle crew to maintain a log, any Command Post or Communications Center should establish a radio log for an incident. This may require assignment of extra personnel to keep the written log for each busy operating position. Completed logs are legal documentation of what was said, when, to whom, and by what means. As such they should be preserved as a permanent part of the incident record.

Communications Procedures

Clear Text Versus Codes

Many organizations use private communications codes, the most common of which are ten codes. Because each agency=s code list is often unique to that agency, the use of codes in a disaster is likely to cause considerable confusion when trying to interface multiple agencies on one disaster radio frequency. Therefore, recommended practice is to discontinue use of any codes during a crisis operation and to only transmit messages in plain language.[63,218]

Use of Ten Codes

Ten Codes are designed to standardize and reduce the length of communications by substituting a code number for commonly used words or phrases. A Ten Code can be used as either a statement a question. For example, "10-20" may be the question "what is your location?" or part of the statement "10-20 intersection Broad and Highland" as "my location is ..." Ten Codes should reduce communications volume; saying "what is your 10-20" defeats the purpose of the code. Although there are a wide variety of Ten Codes, Table 7-1 shows the suggested standard codes developed by the Associated Public Safety Communications Officers:

Procedural Words and Communications Brevity

Certain words have become standard terms used to convey more information and reduce the amount of on-the-air conversation needed. These include the words shown in Table 7-2.

Table 7-1. Associated Public Safety Communications
Officers Ten Code[143,224]

10-1	signal weak	10-20	location
10-2	signal good	10-21	call ____ by phone
10-3	stop transmitting	10-22	disregard
10-4	affirmative	10-23	arrived at scene
10-5	relay (to)	10-24	assignment complete
10-6	busy	10-25	report to (meet) ____
10-7	out of service	10-26	estimated arrival time
10-8	in service	10-27	license/permit
10-9	say again		information
10-10	negative	10-28	ownership
10-11	____ on duty		information
10-12	standby (stop)	10-29	records check
10-13	existing conditions	10-30	danger/caution
10-14	message/information	10-31	pick up
10-15	message delivered	10-32	____ units needed
10-16	reply to message		(specify)
10-17	in route	10-33	help me quick
10-18	urgent	10-34	time
10-19	(in) contact		

Some combinations of words gain general acceptance because they sound official and professional, even though the effect is the opposite. In a disaster, cluttering a radio frequency with these meaningless phrases wastes time and invites confusion. Bad radio practice includes all of the phrases shown in Table 7-3.

Table 7-2. Common Procedural Words[187,243]

Word:	Meaning:
roger	I have received and understand your last transmission
wilco	I have received, understand, and will comply with the instructions in your last transmission
affirmative	yes
negative	no
over	I have completed my transmission to you and am waiting for your reply
out	I have completed my transmission to you and do not expect a reply
I spell	the following is the spelling of a name, address, or other word using the phonetic alphabet
figures	the following words should be written as numbers
wait	standby, I have to delay my transmission briefly
go ahead	I acknowledge your call, transmit your message
clear	meaning the same as "out"
traffic	formal radio message

Phonetic Alphabet

Phonetic alphabets are valuable for spelling words with unusual spellings or for ensuring that a critical word is understood during conditions of poor reception. There are two phonetic alphabets in general use, the International Phonetic Alphabet (adopted by the International Telecommunications Union)[178,224] and the Public Safety

Alphabet (although use of this alphabet is decreasing). The two phonetic alphabets are shown in Table 7-4.

Table 7-3. Bad Radio Practice

Phrase:	Comment:
Be advised that ...	three words that mean nothing - if it is transmitted by radio, the person receiving the transmission has the advice and does not need to be told that it is advice
Roger, wilco	"wilco" (I have received your message and will comply) implies "roger" (I have received your message), so why repeat it?
Over and out	a contradictory message – "over" means that the transmission has been completed and the operator is waiting for a reply but "out" means that no reply is expected
Please ... Thank you ... Sir ...	a radio station has no feelings or rank - good radio manners mean being as short, to the point, and abrupt as possible to avoid wasting time that other stations could use to transmit vital information

Time

Time in communications is universally indicated by 24 hour clock time. Times prior to noon are indicated by the number of the hour as two digits followed by two digits for the minutes - 9:32 am is 0932 and 11:45 am is 1145. Times

153

after noon are indicated by adding 12 hours to the clock time - this 1:30 pm is 1330 and 4:20 pm is 1620.

Table 7-4. Phonetic Alphabets[143,178,188,224]

Letter:	International:	Public-Safety:
A	Alfa	Adam
B	Bravo	Boy
C	Charlie	Charles
D	Delta	David
E	Echo	Edward
F	Foxtrot	Frank
G	Golf	George
H	Hotel	Henry
I	India	Ida
J	Juliet	John
K	Kilo	King
L	Lima	Lincoln
M	Mike	Mary
N	November	Nora
O	Oscar	Ocean
P	Papa (pah-pah)	Paul
Q	Quebec (kay-bec)	Queen
R	Romeo	Robert
S	Sierra	Sam
T	Tango	Tom
U	Uniform	Union
V	Victor	Victor
W	Whiskey	William
X	Xray	Xray
Y	Yankee	Young
Z	Zulu	Zebra

Time in amateur radio[115] and federal government and military messages is often Zulu time (Z) or Universal Coordinated Time (UTC). For east coast locations Z is always 5 hours later than Eastern Standard Time and 4 hours later than Eastern Daylight Time. Thus 1400 hours local in the Eastern time zone in July is 1800 Z. It is good practice to note whether logs are kept in Local or Zulu time and whether times in messages are Local or Zulu time. Because Zulu time is ahead of times in the United States, the Zulu day for communications will change at least 4 hours before the local day ends (based on time zone).

Date-time groups are commonly used in communications to record date and time a message is handled or to identify messages. A date-time group consists of 6 numbers, an abbreviation for the month, and two numbers. The first two numbers (of the 6) are the day, the next four the time in 24 hour clock time, and the last two numbers after the month are the last two digits of the year. Thus 312230 Dec 99 is the date-time group for 10:30 pm on December 31, 1999.

Net Procedures

Amateur radio and governmental high frequency communications often use formal net procedures to control the flow of emergency messages. A Net Control Station is assigned the duty of controlling the net, and stations entering the net call Net Control to check-in, to list any message traffic, and to request permission to communicate with other stations. Net Control tracks the message traffic listed for transmission and directs stations to pass their messages in order of precedence. Stations do not conduct business with other stations without being cleared to do so by Net Control.

Stations remain on the frequency until cleared by Net Control to leave. Although this system may seem rigid and cumbersome, an experienced Net Control Station can ensure that a very large number of messages are passed very rapidly in their relative order of importance.[187]

Individual Alerting

The computer age has created an expectation that a message committed to a computer by e-mail, a fax, a pager, or a voice mail system will be delivered instantly. If it is sent, the assumption is that the other person must receive it. However, that is not true. People often check their e-mail on a set schedule, pager batteries run low, no one looks in the hopper of the fax machine, etc. Any positive control alerting system must require either direct contact with a responsible person or that the individual or agency alerted call the originator of the alert to confirm its receipt.

8. INTERFACE BETWEEN OPERATIONS CENTERS AND MANAGEMENT STRUCTURES

Field and Emergency Operations Centers Structures

As discussed previously, the standard organization for field response is one of the many variants of Incident Command System. Emergency Operations Centers may be organized in a variety of different ways, including the traditional four staff section structure, an Incident Command System, by Emergency Support Functions, or as department and agency representatives. In addition, the presence of Area Command Authorities or a Multi-Agency Coordination System further complicates the command relationships. Table 8-1 summarizes the key characteristics of each of these systems. This variety makes it difficult to easily identify interfaces between the field and the Emergency Operations Center that will apply to all organizations. However, the important issue is whether an interface can be identified for a specific situation.

Points of Interface

Interface Between Command Centers

The need for interface is defined by each element's requirements (see Table 8-2). The Incident Command System primarily needs support. The Emergency Operations Center needs information. These two sets of needs establish the importance of regular communications between the Incident Command Post and the Emergency Operations Center, and the routing of these communications to the appropriate staff officers.

Table 8-1. Command Structure Characteristics[164,167,184]

Command Element	Location	Level	Role	Works With
Incident Command System (ICS)	single scene	tactical	directs scene response	up: ACA or MACS, EOC
Unified Command System (UCS)	single scene	tactical	coordinate scene response	up: ACA or MACS, EOC
Area Command Authority (ACA)	multiple scenes	strategic	coordinate multiple scene responses	down: ICS or UCS up: MACS, EOC
Multi-Agency Coordination System (MACS)	off-site	strategic	coordinate resources and support	down: ICS or UCS, ACA up: EOC
Emergency Operations Center (EOC)	off-site	strategic and policy	direct all locality resources and set policy	ICS or UCS, ACA or MACS, locality

Table 8-2. Typical Incident Command System and
Emergency Operations Center Needs

Incident Command System:	Emergency Operations Center:
1. access to additional resources	1. information for situation assessment and resource prioritization
2. supplies for prolonged operations	2. progress reports for mission management
3. specialized support not included in the initial response	3. level of activity information for disaster declaration
4. alerting and public information dissemination for the general public	4. estimated times of control for resource requirements

Theoretically, the organizational structure in the Incident Command System and the Emergency Operations Center do not have to be similar as long as each structure understands who the counterparts are in the other command center. To some degree, this is simplified if both the Incident and the Emergency Operations Center are organized on an Incident Command System model. The Logistics Officer on scene knows that he or she has to contact the Logistics Officer in the Emergency Operations Center. However, equivalent points of interface can be identified for all of the organizational structures (see Table 8-3).

Lateral Interface

The tendency is for local jurisdictions to be fully occupied with managing their own disasters. However, the reality is that in many disasters the effects will be felt regionally rather than just by one jurisdiction. Active lateral

Table 8-3. Equivalent Functions in Organizational Models

Incident Command System:	Traditional EOC Model:	EOC Organized by ESF:	EOC Organized by Agencies:
Incident Commander	Policy Group	senior executive	senior executive
Planning	Disaster Analysis and Coordination Group	ESF-5	Planning Department
Operations	Operations Group	ESF-3, ESF-4, ESF-6, ESF-8, ESF-9, ESF-10	Police Department, Fire Department, Emergency Medical Services, Public Works
Logistics	Resources Group	ESF-1, ESF-2, ESF-7, ESF-11, ESF-12	Transportation, General Services, Utilities
Finance and Administration			Finance, Administration

(horizontal) coordination between the staffs of Emergency Operations Centers offers the potential for a better understanding of disaster impacts, pooling of mutual aid resources, and coordinated responses to problems that cross political or organizational boundaries. This suggests that horizontal coordination between staff officers with the same roles and responsibilities in adjoining jurisdictions is a wise practice.

Local, State, and Federal Vertical Interface

One of the most common deficiencies in disaster response is the inadequacy of the flow of situation information vertically in the disaster response system. Within a jurisdiction it is not unusual for corporations and major public utilities (including hospitals) to provide only limited or no information on disaster impacts on their organizations to the jurisdiction Emergency Operations Center. Jurisdictions often provide only limited information to state Emergency Operations Centers. They may also provide very little, if any, information to organizations not physically present in the jurisdiction emergency operations center. As a result, state officials may manage by responding to specific calls for assistance without a full understanding of the course of events in the disaster. Vertical flow of information from local jurisdictions to states and to organizations within the jurisdiction is vital to effective response.

At the same time, state government serves as the mechanism for requesting and coordinating Federal government response to disaster events. Local jurisdiction requests for assistance are first addressed with state resources, and should only be passed to Federal agencies if the state lacks the capability to address them. Generally state response is faster than Federal response, and resources committed are generally more familiar with local conditions and commonly accepted practices. If Federal resources are committed to assist a locality, it is vital that local officials keep the state informed as to the progress of the response so that additional resources can be requested or resources not needed can be released.

Interface Procedures

General Procedures

Effective interface between command centers depends on retaining means of communications that can reliably handle large volumes of information (the telephone or the Internet - theoretically it is possible to do this through radio, but the throughput is too slow to be effective). It also depends on finding methods of communications that allow large amounts of information to be captured and summarized in a brief form. Use of voice or fax templates, color coding, standard readiness states or levels of damage, and standard formats for reports ensures that as much information as possible is passed rapidly and in the same way each time.

Effective interface requires regular contact throughout an incident. Even "no change" or "operations normal" reports are important to understanding the big picture. Whether this contact should be hourly, once per shift, once a day, etc. depends on the type of interface needed, the size of the organizations involved, the amount of information in the full system, and the complexity of the event. A general rule of thumb is to push information - to provide the information needed when it is needed and before it has to be requested. The goal is to avoid both information underload and information overload.[197]

In disasters, information flow requires the ability to maintain a functioning system in the face of two challenges, damage and the need to work with a variety of partners. Experience in World War II demonstrated that Civil Defense communications between system components could provide required information within reasonable time limits even

162

under the constraints of significant disruption.[228,273] Evaluation of data flow in large and complex search and rescue incidents shows that technical communications problems are not the primary causes of communications failures – the actual deficiency is in interagency coordination.[73]

Specific Types of Information Needed

Specific information requirements depend on the types of organizations, types of disasters, and general practice in the jurisdiction. However, the following types of information are some of those commonly needed:

... time and severity of impact of disaster effects
... injuries and fatalities
... shortfalls in resources and expected needs for resources and supplies
... impacts to supporting infrastructure
... types of resources sent and expected arrival times
... critical outages and expected time of return to operations for systems
... population protection measures recommended for implementation, or in place, or being phased out
... overall capability assessments
... psychological, social, and economic impacts
... the time of disaster declaration, activation of the Emergency Operations Plan, and opening of the Emergency Operations Center
... time and location resources were committed to response
... automatic aid and mutual aid response and aid compact activation.

Coordinating Information Flow

One of the most difficult tasks in either a field Command Post or an Emergency Operations Center is management of the flow of information. Internally, routing slips and forms designed with check-off boxes to track who has seen, or not seen, the information on the form offer one solution. Multiple form copies on paper printed in a variety of colors, with a fixed distribution for each color provides another, less flexible solution.[245] Even if these fail, periodic briefings on a set schedule during a shift can provide a reasonable assurance that key information will be seen or heard by the individuals who need it.[176,181] In addition, rigorous paper control, including such measures as banning the use of scratch paper and requiring that all forms be filed on a regular basis, reduces the chance of misplacing information.

However, the flow of information to other Centers is not so easy to manage. To encourage information flow, internal message forms should include a check-off block to indicate that data should be passed to other Centers. It may be worthwhile to assign a liaison position specifically to manage information flow. It may also be wise to maintain a log of information passed vertically or laterally and to whom the information was transmitted.

Federal Emergency Management Structure

Four Phases of Emergency Management

The Federal Emergency Management Agency uses a four phase life cycle model to describe programmatic actions taken to manage disasters.[117,156,256]

Mitigation

Actions taken prior to a disaster to prevent its occurrence, or to reduce disaster effects if there is an occurrence, are termed mitigation. Typical examples include land use restrictions in flood plains (which ensure that a flood will not impact the built environment), building codes in high wind areas (which reduce the likelihood of damage to the envelope of buildings), and insurance (which reduces the economic impact of a disaster).

Preparedness

Preparedness actions are those which prepare an organization to respond effectively to a disaster to limit human impact and damage to property. Examples include training, disaster exercises, planning, developing Standard Operating Procedures, and equipping rescue forces with appropriate vehicles and tools.

Response

Response actions are taken to provide emergency services during the actual impact of a disaster and to start the recovery process. Examples include lifesaving and rescue activities, fire suppression, immediate restoration of utilities, and initial damage and needs assessment.

Recovery

Recovery actions are those taken to attempt to restore the community to normalcy. This does not necessarily mean restoring the community to its pre-disaster state; in a catastrophic event that may be impossible due to the cost and the damage sustained. However, it does mean restoring basic lifelines and providing the framework for governmental services and economic and social recovery. In international disaster response, recovery may be used as the framework for actual development work. Any recovery should be linked to mitigation activity to reduce the threat from future disasters.

Crisis Management - Consequence Management

Introduced as a result of planning for incidents of terrorism, crisis management and consequence management have started to become standard terms for two phases of disaster response. Crisis management in this context refers to law enforcement actions taken to identify and apprehend perpetrators of terrorist acts. Consequence management refers to all actions, normally coordinated by emergency management officials, to deal with the impact of those acts.[249] Consequence management is increasingly being used in other contexts to describe response and recovery.

Note that the terrorism program use of crisis management is distinctly different from the commonly accepted definition of a crisis as a critical turning point in a major event.[93] Crisis management is more properly the management of the critical point in a major emergency or disaster, normally falling during the range of time period from immediate pre-impact, through impact, and normally terminating during response.

The Federal Emergency Support Functions

The Federal Response Plan establishes twelve Emergency Support Functions (or ESFs).[249] An ESF is a major function that must be performed during a typical disaster, organized with a lead Federal agency as coordinator of activity and with other supporting agencies assigned to provide resources. Emergency Support Functions do not address all possible areas of disaster response; they only address those areas in which the Federal Government has significant capabilities. Emergency Support Functions are as listed in Table 8-4, below.

Table 8-4. Emergency Support Functions[249]

Function Number:	Function Name:
ESF 1	Transportation
ESF 2	Communications
ESF 3	Public Works and Engineering
ESF 4	Firefighting
ESF 5	Information and Planning
ESF 6	Mass Care
ESF 7	Resource Support
ESF 8	Health and Medical
ESF 9	Urban Search and Rescue
ESF 10	Hazardous Materials
ESF 11	Food
ESF 12	Energy

It is important to note that some states have adopted additional emergency support functions within their state response structure. Florida, for example, has ESF-13 through ESF-17 to provide management structures for such issues as law enforcement, volunteers and donations, and animal issues.[99]

Federal Response

The initial response to disasters is always a local response by a community's first responder agencies (fire, law enforcement, emergency medical services, emergency management). The second response is normally a state level response, using the resources of the various departments of state government, including law enforcement, public health, transportation, forestry, and the National Guard. Note that the National Guard is typically activated in its state mission, rather than as a Federal resource, preserving its capability to perform law enforcement roles. Federal government response to disasters is the third level of response and will normally start to arrive days, rather than hours after the event.

During the response phase the Federal Emergency Management Agency will dispatch an Emergency Response Team – Advanced Element (ERT-A) to the state Emergency Operations Center to assist in coordinating requests for federal assistance. This will be followed by the full Emergency Response Team (ERT). The ERT, in conjunction with state officials, will conduct damage assessment and establish a Disaster Field Office and Disaster Recovery Centers as the start of the recovery phase.[249] A common misunderstanding of Federal response is that military resources will be immediately dispatched to a disaster scene. United States Armed Forces roles are limited, and units from bases close to the impact area may be otherwise committed. A Request For Federal Assistance process identifies needs and sends the most appropriate resource to provide the needed capability.

9. OPERATIONAL RISK AND RESOURCE SAFETY

Framework of Risk

Major emergencies and disasters are inherently dangerous working environments. Emergency response personnel are working under the same conditions that have endangered property and the lives of citizens. Operationally there is increased potential for injury to or the death of response personnel and the damage or destruction of expensive and operationally valuable equipment and vehicles. From a program management standpoint there are the costs associated with personnel and property loss, legal liabilities, and the loss of time.[124]

These risks can be controlled to a degree by a variety of measures, including training, proper equipment, Standard Operating Procedures, use of an Incident Command System or other management structure, safety programs, accident investigations, and a variety of approaches to transferring costs, including insurance.[124] However, these measures are not incident specific and sometimes provide only limited guidance for decisions in the incident Command Post or Emergency Operations Center.

Risk Defined and Assessed

In emergency management the commonly accepted operational definition of risk is that it is the product of the possibility of an event occurring times the impact if it does happen. This relationship can be diagrammed as shown in Figure 9-1.

Figure 9-1. Risk Relationships

Probability - *Low* Impact – *High* RISK – *MODERATE*	Probability – *High* Impact – *High* RISK - *HIGH*
Probability – *Low* Impact – *Low* RISK - *LOW*	Probability – *High* Impact – *Low* RISK – *MODERATE*

Operationally these assessments are often made based on individual managers' experiences and subjective evaluations of conditions. However, it is possible to quantify specific elements of the assessment and develop a scoring system that provides managers a reference point for decisions. One example used for assessing risk in flight operations included:

… available planning guidance
… aircrew rest
… weather and light
… aircrew experience
… terrain.[42]

Similar lists of risk assessment factors can be developed for other types of operations, tailored to the operational environment of the organization.

A quantifiable assessment has the advantage of requiring the manager to examine known factors that increase risk and account for them. In addition, if used as a worksheet, it provides documentation that these factors were considered and found to be acceptable.

Risk Versus Gain

Almost any action directed during a disaster carries with it some element of risk. This suggests that every assignment should be based on careful consideration of the balance between risk and the potential gain that may come from the assignment.[57] One approach to this is summarized in Table 9-1.

Table 9-1. Risk Versus Gain Comparison[57]

Requirement	Potential Gain	Risk Accepted
structured plan of action	life saved	significant
structured plan of action	valuable property saved	minor
not applicable	life already lost (body recovery) or property that is not savable	none

It is important to note that this comparison model addresses significant risk. This is different from the combination of a high probability of loss of rescuer life and a low probability of a successful rescue that may confront emergency managers in extreme conditions. During some situations (for example, the storm surge period during hurricane landfall) suspension of all emergency response may be essential to avoid loss of rescuers and vehicles and equipment essential to response after the immediate impact passes.

It is also important to note that the degree of risk in most tactical situations may be sensitive to two factors. First, the size and complexity of an incident has been shown in one research study to have a direct relationship to numbers of injuries sustained by fire department personnel assigned to control it. Second, the number of personnel assigned to respond to an event may be inversely related to injuries sustained. In simple terms, more complex incidents equal more injuries, and fewer responders equal more injuries. Therefore, a key risk control measure must be to assign sufficient resources to any incident.[265]

Urgency

A complicating factor in determining whether to commit forces to a risky situation is urgency. Some situations are obviously urgent – a response is needed now if there is going to be a response, and if the responders are to have a chance of saving life. Other situations may be less urgent, allowing either action to be deferred until conditions improve or a more measured approach to be taken to gather more information and prepare a well-designed plan. The search and rescue community has done extensive work on evaluating urgency, and the factors considered in urgency determination for a search and rescue response may offer perspectives for disaster response:

    … missing person's age
    … missing person's medical condition
    … number of lost persons
    … weather
    … equipment carried by the missing person
    … the missing person's experience
    … area terrain and existing hazards

... the history of previous incidents in the area.[207]

Monitoring Resource Status

The range of actions needed to assure safety on an incident scene are extensive, and require the services of an incident Safety Officer. At the Emergency Operations Center level safety remains an equal concern, but the areas in which intervention is possible are more restricted. In addition to risk assessment prior to committing resources, monitoring the status of resources committed and maintaining accountability are key safety procedures.

Accountability Methods

Accountability of resources on scene at a major emergency incident depends on three components: accurate tracking of resource assignments, regular accountability checks, and tracking to the level of individuals. Tracking of resources requires use of tactical worksheets, regular reports from units as to their location and assignment, and the assignment of resource tracking to an individual assigned as an accountability manager.[198]

To continuously monitor accountability during on-scene operations, checks should be initiated at scheduled intervals. At the unit level, each unit leader on scene accounts for all personnel assigned to the unit. Radio reports by units serve to account for all units. If a unit does not report, immediate action should be taken to locate the unit and confirm its safety. Who initiates this process depends on local Standard Operating Procedures. One fire department requires the Dispatch Center to initiate an accountability check with the Incident Commander 20 minutes after the first unit arrives on scene.[203]   This time period was selected based on department experience with the flow of operations and limitations of equipment to ensure that accountability would be established starting from a point of highest potential payoff in maintaining safety.

A variety of systems provide accountability of the relationship between people and units. These include Velcro based Passport systems with plastic tags for individuals, and paddles for each unit. Other systems use clip rings for each individual or bar code reader systems.[47]   Magnetic board systems based on tactical worksheets can be used to track assignments and locations of units.[199]   The advantage of these systems lies in the ability to track individual assignments and know at any given time which individual should be assigned to which unit and what task.

Importance of Scheduled Accountability In Route

In large-scale disasters, unit times from dispatch to on-scene may be extended, especially if these resources have significant distances to travel or are moving through disaster impact areas. This means that a process of trip following must be initiated as soon as units depart their home stations in route for their assignments and continue until they arrive on scene. Procedures in air search and rescue require an operations normal report by search aircraft every 15 minutes and a position report every 30 minutes. Failure to report initiates a search for the search and rescue aircraft on the assumption that it may have experienced a system failure or other emergency condition and be in need of assistance.[242]   A similar process should be required for units moving by road, although an hourly contact schedule is sufficient for most conditions. In this model, the Emergency Operations Center or Dispatch Center remains responsible for units until they are under the control of the site Incident Commander. When the unit is released to return to its home station, the same procedure is followed until safe arrival is confirmed.

10. INFORMATION MANAGEMENT AND DISPLAY

Role of Information

Disasters generate huge amounts of data (individual bits of knowledge about single things), ranging from insignificant to critical in value. For this data to be of value in bringing an incident to the best possible resolution, a number of things must happen:

(1)   It must flow.   In a changing situation, data is perishable and must be rapidly transmitted to those who need it, both horizontally and vertically.

(2)   It must be evaluated and compiled into information. This requires data be collated, associated with like data, and assessed for quality and trustworthiness to paint a picture of the situation.

(3)   It must be made available to decision makers in a way that facilitates making the correct decisions.   This requires reports that highlight information of strategic and tactical value, display systems that show the relationships in significant   problems,   and   assessment   methods   that communicate the confidence level assigned to information and the uncertainty associated with absent information.

(4)   There must be adequate technical infrastructure to support the process, and that infrastructure must be used. Information management infrastructures do not have to be more complicated than wall mounted status boards, paper logs, and telephones.   However, in general, the simpler the infrastructure the smaller the data load it can handle.

Unfortunately, when a complex technical infrastructure of databases, computers, and communications is in place, it often remains unused.[142]

(5)   Adequate trained staffing must be available to make the process work.   This requires a significant number of administrative personnel to manage data entry and system operation.[142]   Training remains a critical shortfall – research suggests that staff of Emergency Operations Centers have a very low level of proficiency in installed computer data management systems.[100]

(6)   And the entire process must present the correct picture in the right amount of detail and with high reliability. The trend today is to generate data in fine detail using Geographic Information Systems, large databases, Global Positioning Satellite location data, etc.   Three concerns emerge from this.   First, information overload is a significant problem – the amount of information and the fineness of its detail   may   actually   inhibit   effective   decision-making.[197] Second, data reliability is potentially an even greater problem.   Large databases require constant, daily staff effort for their maintenance, and the reality of staffing is that the level of effort available often is insufficient to maintain the accuracy of data after its initial entry.   Third, data collected may be of such detail[54] as to have no practical application in dealing with real disaster problems.

Use of Computers

General Use of Emergency Management Software

A   variety   of   specialized   incident   command,[137] emergency management, and business continuity software is available to manage information and make it available for decision making. Programs vary in sophistication from very simple task tracking software to fully integrated systems that include mapping, geolocation, and a full range of text tools for management. An examination of the features of commonly used software packages (EIS/GEM, EM2000, E-Team, and Softrisk) suggests typical uses of emergency management software include:

(1)   Incident log. Incident logs allow information to be entered in essentially the same way as a paper log, except that the data is stored by the computer. Typically such logs will automatically time stamp and number the entry. Because the log form is designed to include significantly more options for entering information (blocks may be included for name of the person contacted, telephone number, organization, status of the action, and subject), the entries may better capture the complete action than an equivalent paper log.

(2)   Task tracking and prioritization. Task tracking allows creation of an entry for each assigned task and periodically updates that entry as work is done to resolve the problem. With prioritization, those tasks that are most important are highlighted by the software. In some systems, tasks that have not been assigned or that are overdue for completion may flash or otherwise be highlighted.

(3)   Incident geolocation and situation display. The inclusion of mapping software allows incident sites to be represented on the map with an icon, and that icon to be related to a log entry. Clicking on the icon may refer the map user to the log entry. The capability to draw on, letter, or add symbols to the map allows you to display a range of

additional situation information (such as plume data from chemical releases, flooded areas, road blockages, etc.). A high degree of geographical accuracy may be possible, with the ability to measure distances accurately.

(4)   Resource inventory tracking, and geolocation. A resource inventory function allows resources to be established in a database, along with all needed details, those resources to be assigned a standard icon, and that icon to be geolocated on a map to indicate the resource's normal station. Resource tracking functions allow determination of the quantity of resources committed to tasks, and may allow identification of resource by task. Resources may include key staff, with personnel records, and with the capability to assign them to specific duties, with an automatically generated log entry to document the assignment.

(5)   Plan and checklist execution. Most software will allow the Emergency Operations Plan to be stored as a file so that it is electronically available. This feature is more useful in theory than in fact for most applications. However, the inclusion of Checklists can be more useful. Check off of a Checklist item may generate a log entry automatically indicating its completion.

(6)   Situation Report preparation. Some software generates a template for a standard Situation Report, and may generate entries automatically.

(7)   Resources requesting and dispatching functions. With geolocated resources, it may be possible to deploy an icon on the map in conjunction with resource tracking functions. This may generate an automatic log entry and count down the number of resources available.

(8)	Cost tracking. Resources may include a cost entry.	Software may include a capability to estimate the cost of response operations based on the resources committed and the time that they are deployed.

(9)	Communications and data exchange. The software may include the ability to exchange information and messages with other systems, as e-mail messages, proprietary messaging systems, packet data transmission, or through local or wide area network exchange of data. Networked systems offer the ability for assignment of taskings from position to position, reporting of results based on the assignments, and the sharing of a wide variety of information among all command center staff. The potential for exchange of data between Emergency Operations Centers laterally or vertically also exists if all Centers are using the same software or at least the same message protocols.

(10)	Form production. Software can print standard forms and T Cards as record copies.[137]

Geographic Information System

Mapping systems integrating Geographic Information System databases provide very sophisticated capabilities for Emergency Operations Centers. At the most obvious level, availability of high quality maps through computer displays supports planning and situation analysis. However, the ability to overlay a wide variety of demographic data, resource locations, flood plain information, shelters, evacuation routes, etc. in whatever combination is desired for decision making provides a significant enhancement. Finally, the use of Global Positioning System verified field reports allows overlay of current disaster effects.[131]

Electronic Mail

The use of electronic mail in day-to-day operations is nearly universal in government and business. However, electronic mail use in a disaster should not necessarily be conducted in the same way as in a normal business environment. Because e-mail in a disaster becomes part of the disaster documentation, users of e-mail should follow certain precautions:

(1)   Clearly identify for whom the message is intended. A command center account may be shared by a number of persons or functions. Use of a TO or FOR line before the text is a good practice. If the message is going to a function, indicate that function. Do not address a message intended for the function to an individual; if he or she has gone off shift, the recipient may hold it for them without reading it or taking action. And several years later, when trying to reconstruct what happened, readers will have no idea whom is indicated by the e-mail address.

(2)   Check the clock that time stamps messages. If there is any doubt about its accuracy, clearly indicate the date and time the message was composed in the text.

(3)   E-mail messages may be read by anyone and forwarded to anyone else. Do not make disparaging comments about victims, other agencies, etc. Personal opinions in e-mail become part of the official record and may become part of the public record. They are certainly discoverable in litigation, and may be subject to Freedom of Information Act disclosure.

(4)   Do not violate individual privacy in messages. If names, medical conditions, loss, or fatality data must be transmitted by e-mail, mark messages clearly as including confidential material.   The   same   caveat   applies   to transmission   of   information   that   is   proprietary   business information.   This will not stop someone from misusing it, but it may offer some grounds for legal or disciplinary action against them.   Include   the   word   CONFIDENTIAL   or PROPRIETARY in the subject line so that it is immediately visible in the e-mail program message list.

(5)   If a message is high priority, indicate so using the   same   precedences   used   for   internal   or   radio   message handling.   Include the precedence in the e-mail subject line.

(6)   Sign all messages by ending the message with the author's name, as normally signed, and job title.   Do not assume that anyone knows the identity of the writer based on the e-mail address.

(7)   If the message is an exercise message, start and end the text with the word EXERCISE.

(8)   If   forwarding   a   standard   report,   mission tasking,   or   other   document   that   regularly   comes   in   a common format, try to include the information in the e-mail in the same format as the document would be if word processed.   It will make it easier to read and understand.

(9)   Spell check messages.   It is possible to avoid embarrassment   from   a   very   few   errors   in   handwritten documents, but most good e-mail programs now include a spell checker.   Just as is the case for other documentation,[150] the assumption is that sloppy e-mail messages using poor

grammar, and with multiple spelling errors, reflect the performance of sloppy work.

The Internet

The Internet provides a highly distributed and survivable source for a wide variety of information services. Most Federal government agencies with an information role, including in such areas as weather, seismology, and wildfire, provide access to their services through Internet access. State emergency management agencies generally provide access to current Situation Reports and other data through their sites. This allows a local emergency manager or business continuity professional to access high quality information in near real time during an event.

There are concerns of which potential Internet users need to be aware. First, not all Internet sites are created equal, and the quality of information in many sites that appear to be authoritative is open to question. Site users need to determine the identity of the site owner and establish the site's credibility. Second, sites may not be automatically updated. Information currency is a significant issue, and users should review notices of most recent updates carefully. Third, information overload, or the more insidious information arrogance phenomena, are real possibilities. Much of the information available requires a certain degree of education in the underlying science beyond that of high school basic science courses. It is easy to fall into the trap of assuming that access to a large volume of information creates instant expertise in the field.

The Internet offers the potential for management of emergencies remotely. Already The Virtual Emergency

Operations Center, a volunteer organization, is performing distributed Emergency Operations Center functions for its served agencies on the Internet.[95] As technology evolves, emergency managers and business continuity professionals should continue to explore ways in which the Internet can provide communications, data sharing, reporting, and task management functions.

Managing Information Flow

Information is the currency of Emergency Operations Centers. In emergencies, knowledge is power, and that power is multiplied when the knowledge is shared with everyone else who needs to know it. This places a high value on the acquisition of information, its analysis, and rapid dissemination to all other sections of the command center that need the information.

Information dissemination within an Emergency Operations Center traditionally has been handled by the use of an incoming Message Center that receives all messages, records them on multiple sheet forms, and distributes the red, yellow, white, green, etc. colored copies according to a set scheme for each type of message. One copy goes to the primary section, one copy to the Center manager, one copy to be posted on the status board, one copy stays in the Message Center record, etc. The actual scheme is typically set by the Center's Standard Operating Procedure.[245]

This approach has the advantage of providing multiple documentation of the information and providing a hard copy rapidly to everyone who is potentially interested. In addition, the Emergency Operations Center supervisor has a copy of everything to ensure he or she is fully informed of everything that is happening in the Center. The danger is in the generation of huge amounts of paper and the swamping of key positions with too much information. Individuals who, in the normal course of work, receive twenty pieces of paper a day for decision-making are rapidly overwhelmed when they receive that quantity in an hour. In addition, the representation of each item of information by several pieces of paper creates difficulty in tracking completion. Finally, problems which come directly to staff sections from the outside may not be recorded and distributed.

An alternative approach is to place each piece of information on a single sheet of paper that is initiated by whomever receives the input. This paper is then routed in sequence to those who need to see it, based on the judgment of the staff officers handling the problem. Actions taken are documented on the single sheet so that the one page represents the history of the problem. The supervisor sees the problem at the start of the process and when it is completed.[52]

The advantage of this approach is that the volume of paper is reduced and that everything concerning the issue is in one place. A single notebook can be maintained that oncoming staff members can check to see the entire history of the incident. And each person who handles the information knows what has been done previously on the issue it represents. Each person who is aware of the information is documented by use of a check-off column.

However, because there is no central repository, problems can be pigeonholed in one section's in-basket or just simply lost (surprisingly enough, experience suggests lost messages also happens with multiple copies). This approach requires that other methods, such as computer-based task tracking or wall displays, be used to identify all of the issues being worked — the supervisor no longer has a pile of copies that he or she can use to control operations. The single page system requires a more highly trained staff and probably works best in smaller centers.

The issue of documenting who is aware of what information, when, is important, not only to eliminate questions as to who has read a document, but also for after action review and legal documentation. It is a good practice for forms containing information to either have a read-and-initial block or for persons reading a document to indicate so with their initials and a date-time group in the margin.[56]

Internal Message Forms

All information received in an emergency operations center should be documented by recording it in writing, either on a paper form or in a computer log entry (most emergency management computer softwares allow log entries to be sent as messages to staff sections interested in their contents). Paper message forms vary widely based on the standard operating procedures of centers, but may include the following:[37,50,78,215,245,254]

    ... PRECEDENCE to indicate urgency of the problem (to either the sender or receiver)
    ... a FROM and TO section
    ... an INFORMATION address line for delivery to other interested sections
    ... DATE and TIME of the message
    ... INCIDENT identification (in case of multiple incidents and for filing purposes)

     ... message SUBJECT
     ... TEXT of the information
     ... originator's SIGNATURE
     ... a box for action taken by the recipient of the message
     ... a read and initial box
     ... a numbering box for page numbering as part of the record

Types and Functions of Status Boards and Maps

The primary means of displaying information so that all in the Command Post or Emergency Operations Center can see and use it for decision making is through the use of wall mounted status boards and maps.[245] Computer software on local area nets offers similar capabilities, but the status board remains important for the flexibility and ease of access it provides.

The functions performed in a command center determine the types of status board and map displays that will be needed. In general, these displays are used for:

(1) Displays of situation information. Incoming reports can be tabulated on a status board as text and displayed on maps to provide spatial relationships. Status board displays will typically indicate time of report, location, and explanation. Specialized types of situation boards may be used for weather, road closure, sheltering, computer or communications system outages, damage,[201] and other situation information.

(2) Task tracking. Generally specific tasks are tracked on status boards showing the time the task is received, the

187

nature of the task, the resources committed to it, and the time the task is completed. Task assignment locations may be placed on a map. In some types of responses, completed tasks may be left on the map to visually identify areas that have not been covered (for example, search area coverage, completion of damage assessment, or reporting of impacts).

(3) Resource tracking. Resources are tracked by status boards which display type of resource, identification, status, and location. Resource current locations may be placed on maps. Personnel availability,[240] scheduling, and job assignment status boards are types of resource tracking.

(4) Shortfall identification. Needs for supplies, equipment, field resources, or support functions may be displayed on a status board that indicates when the need was identified, what is needed, when the need is filled, and what filled it.

As an example of the types of status boards jurisdictions may find useful, Pase Incorporated sells a 17 status board kit for local Emergency Operations Centers. Status boards in that kit include boards for:[179]

> event status
> news media status
> special needs status
> shelter and facility status
> area closings
> evacuation status
> weather
> casualties and damage
> incident command system staffing
> contracts and agreements

hospital bed status
carrier status
route status
resource status
work schedule
personnel status

Maps may perform two functions in a command center. As noted above, they may be used to display information that makes more sense when shown in relationship to the geography of an area. They also serve an important reference function in helping staff members identify key locations, select routes, assess potential impacts, and identify problem locations or components of systems.

To be useful, displays must be able to be seen. This means that the larger the Center in area, the larger the displays and the writing on them must be. If the Center is a small conference table, you can actually meet your display needs with status boards in plastic sleeves mounted on standup notebooks on lazy susans. If, however, the Center is a 40 foot by 40 foot room, people in the back of the room will only be able to use the information if the displays are floor to ceiling in height and the lettering is 5 to 6 inches in height.

To some degree the need for very large displays can be reduced by careful location of the displays. In general, a display should be positioned so that it is in front of the staff section that will primarily use it.[245] For example, a shortfall display is primarily going to be used by the logistics staff and situation displays by planning, analysis, and operations staffs. A status board dedicated to communications circuit status is probably of greatest use to the Communications Center.

189

In general, displays should be located in front of operating positions, not behind them. Although it may be tempting to use every available section of walls for maps, charts, status boards, and whiteboards, putting displays behind work stations forces the staff to turn around to use them. It is easy to overlook information that requires that degree of extra effort.

Maintaining Status Boards and Maps

Displays are useful aids to decision making when the information on them is current and when information is presented in standard ways. Updating of status boards should be a normal part of the handling of any information — it may be worthwhile to include a check-off block for status board posting on message forms to remind staff that the information should be posted.

Status boards and maps should be posted according to a Standard Operating Procedure. Considerations in posting may include the following:

(1)   The use of color coding to indicate priority or seriousness of actions or information.[147]   Thus RED items are of more importance than those posted in BLACK.

(2)   The use of colors on maps to indicate the level of confidence in information plotted on them.   RED might indicate high confidence, BLUE a low confidence level.

(3)   The use of colors on maps to indicate the source or types of resources or the types of damage. For example, all fire units could be posted in RED, all emergency medical

services in BLUE — alternately working fires could be RED, and flooding BLUE.

(4)   The use of standard symbols on maps to indicate disaster impacts, facilities, and response resources.[1,33]

(5)   The use of standard abbreviations on maps and status boards—for example CO for company, BN for battalion, DIV for division, EOC for Emergency Operations Center, CP for Command Post, TF for Task Force, ST for Strike Team, etc.

(6)   Posting of a legend with the maps or status boards so that persons not familiar with the coding and symbols will be able to gain familiarity rapidly.

Maps, in particular, require training on their use. A surprising number of people cannot read a road map, much less a topographic map, or specialized maps of flood plains, land use, coastal waters, utilities, communications circuits, etc. There are some general conventions that apply to almost all maps and that everyone working in a command center should be familiar with:

(1)   Unless otherwise indicated, the top of a map is normally north, the left hand side is west, the right hand side is east, and the bottom of the map is south.

(2)   Roads are normally shown by solid lines of increasing width based on the road quality. Roads are normally identified on maps by markings that are a reasonable facsimile of the road signs used to mark them. Railroads are normally a single line with crossbars representing the railroad ties.

(3)   Water and waterways are normally blue; forested areas are normally green (which includes parks in some cases); cities and urban areas are normally yellow (on road maps) or red (on topographic maps).

(4)   Many maps are overprinted with a grid system. These grids may be set by the map maker based on distance (a half-mile on a side, for example) or by one of a number of standard grid systems including latitude and longitude, Universal Transverse Mercator (military maps), forestry or township and range, National Search and Rescue Grid, etc. With the exception of grids designed by the maker of a particular map, all of them have a subdivision system that allows pinpointing of position within the grid to a reasonable degree of accuracy for the intended use.

(5)   Most maps are printed with a scale and measuring rules.   Small scale maps (1:100,000 or more) cover large areas; large scale maps (1:24,000 for example) cover small areas.

(6)   Maps are not necessarily current.   Almost all maps include a publication date, and in some cases a date that indicates how old the information on the map is.   The explosion of interest in Geographic Information Systems has provided localities the capability of having very accurate maps. However, the data is only as good as the last update.

(7)   Specialized maps tend to omit all detail that is not important to the information they are trying to present. Use of a specialized map for a general purpose is risky as the map may not have information that is important — for example, it is difficult to use a maritime chart as a tool for identifying roads, even coastal roads.

T-Cards

T-Cards are commonly used in wildland firefighting as a visual method of resource control. Each type of resource, including key staff personnel, is entered on a card. Cards are color coded by resource type to enhance rapid resource recognition and sorting.[33]   The color coding used for wildland incidents is as shown in Table 10-1.

Table 10-1. T-Card Color Coding[33]

Card Color:	Resource Type:
rose	Engines
green	Handcrews
yellow	Bulldozers
orange	Fixed-Wing Aircraft
blue	Helicopters
tan	Task Forces and Miscellaneous
white	Personnel
gray	Location Labels

T-Cards for wildland firefighting are supplied with standard blocks for entry of information about a resource, its location, organization, status, etc. For other applications, it may be preferable to have cards printed with organization specific blanks. An example of such a card is shown below in Figure 10-1.

Figure 10-1. Generic T-Card

Agency:	Type:	ID:

Assignment:	Check-In:

Home Station:

Leader:

Crew:

Available Until:

Status:
___ Available   ___ Out of Service
___ Assigned

Remarks:

T-Cards are called T-Cards because of their shape. The upper portion of the card is cut to extend out of a slot in a metal or cloth rack used to hold the cards in a command post or emergency operations center.

Tactical Worksheets

Tactical Worksheets, also known as command boards,[204] are commonly used by Incident Commanders and in the Command Post to assist in the control the early stages of a major incident.[28,101,157,158] The Tactical Worksheet performs a number of key functions. It serves to organize, and provides a place to display, information needed for decision making — large whiteboard or laminated plastic tactical worksheets are particularly useful for information display. The structure of the worksheet serves as a reminder — blank spaces suggest missing information and provide a stimulus to ask for reports. Well-designed worksheets can incorporate standard lists of items that must be done at every incident.[49] Tactical worksheets help in the tracking of the incident situation and available resources, performing an important function in accountability.[47] And they can be photographed or copied to provide incident documentation.[49]

Tactical worksheets are designed to meet the requirements of individual Incident Command System positions and the operational philosophy of the response agency. In general, any tactical worksheet might consider having space for the following elements, as required by the role the worksheet and the duty position are expected to play:

… time last updated
… worksheet duty position
… diagram of the incident scene
… resources, requested, in-route, staged, parked, assigned (and the assignment), released
… hazards
… objectives
… key staff positions

... event impact, damage, casualties
... checklist of key steps

Completion of tactical worksheets requires self-discipline and adherence to the organization's Standard Operating Procedures. Inexperienced supervisors often feel that the worksheet is an unnecessary burden when they should be concentrating all of their time on the incident. In actuality, use of the worksheet helps to organize information and the response and actually speeds up operations. In the stress of a major event the worksheet eliminates the need for recall of information and makes the most current picture available for decision-making.

11. INCIDENT DOCUMENTATION

Standard Documentation

The events during an emergency incident can be recorded for future reference in a number of ways, including on paper, on audio tape, and in an electronic record. The most common approach is the use of paper records, which may vary widely from purpose designed paper forms to bound ledger books to the common legal pad to scraps of paper and yellow stickees. Legal pads and scraps of paper offer the lowest amount of functionality and the highest probability that incident records will either be discarded or filed in a wide variety of desk drawers. Bound ledger books have significant advantages. As a bound book with numbered pages, the record immediately gains credibility – it is very difficult to remove or add something to the record maintained at the time. In addition, the bulk of the ledger reduces the chances that it will be inappropriately filed or lost.

However, purpose designed forms contribute to incident management in a way that a ledger book cannot. The primary advantage of the form is in its ability to assist in the organization of information for decision making, combined with a prompting function that helps the user ask the right questions and record the right data. There are a wide variety of incident forms. The closest thing to a national standard are the forms used by the National Interagency Incident Management System Incident Command System (NIIMS ICS).[181] However, each agency should consider the actual need in making a choice on form design. NIIMS ICS forms are optimized for wildland firefighting. In the environment of the Emergency Operations Center there may well be better ways to collect and display information for emergency operations use (especially in business command centers). Typical types of forms that should be considered for use in any command center include:

... incident logs or journals (used to record the continuing flow of events)
... message forms (used to transmit information to or receive it from others)
... resource forms (used to track the status of resources)

Completing an incident form requires an organized approach to the form. First, determine the correct form for the job. Second, identify what goes in each block of the form – a good practice is to have directions for the use of the form printed on the back of the sheet. Third, record the information as completely as possible. What is not recorded on the form is lost forever – in a busy event it will have disappeared from human memory in 30 minutes. Fourth, do not abbreviate unless everyone understands the abbreviations. Fifth, fill in every block; blocks that are not applicable in this situation should be marked N/A. A filed-in block shows the subject was considered and the individual completing the form made a determination about it; an empty block may indicate no one asked the question or took the action. In general, forms are best completed in black ink – it copies the best, and ink makes any changes obvious.

Computer based incident documentation such as InciNet,[137] EIS/GEM, Softrisk, EM2000, E-Team, etc. assist in documentation by recording and storing the form information in a single database that other users can access. Although each system has its own operating instructions, in general the first step is to open an overall incident journal, and then complete data for each event or change on a separate record within that journal. These are typically dated and time stamped automatically and maintained in sequence. One of the general advantages of computer systems is that most of them are very good at tracking assignment of resources and individual personnel to emergency tasks and to tracking follow-up actions until a specific event is completed.

If computer systems are used to manage disaster records, it is also important that operators are trained on backup procedures and on maintaining records on paper in the event of a failure, and in how to rapidly transition from computer to paper. This transition is helped if forms are designed to resemble the computer system entry screens in use as closely as possible.

The Incident Log

An incident log is a written record of the actions taken by an emergency operations center to resolve a crisis. Logs are legal documents; the presumption is that actions in the log were actually performed, but actions not in the log were not performed.

What goes in a log? The best answer is everything; the log should be a complete record of everything that happens in the Emergency Operations Center, including:[237,239]

… who is on duty
… when shift changes occur, and the fact that a shift changeover briefing is given
… visitors to the Emergency Operations Center
… information received – what, who, when, where, why, and how

... changes in resource status

... dispatching of resources on specific assignments and the results of those actions

... warnings given and protective actions ordered

... press releases

... equipment outages and return to operations

A common shortcoming in logs is the failure to document thought as well as action. Emergency Operations Centers exist to make decisions, and the actions resulting from decisions are typically recorded in the log. However, long after the fact it is often difficult to reconstruct why decisions were made. Actions that made perfect sense at the time may seem completely nonsensical with the benefit of time and hindsight. Therefore, whenever decisions are made, it is important to document, concisely but thoroughly, the factors that resulted in the decision. This is an important defense against suggestions that decisions were made negligently or capriciously.

Should each duty position in the Emergency Operations Center maintain its own log? To a degree this depends on the type of work and the method of its accomplishment in the Emergency Operations Center; it also depends on the technology used to support operations. In an Emergency Operations Center that focuses on a relatively narrow range of actions, and in which all functions tend to be involved in working in a coordinated manner on specific tasks, a single log has significant advantages. One single document reflects the total range of actions completed in the Emergency Operations Center. Duplication of effort is reduced, as information is recorded one time in one place. And there is less chance that important information is scattered among a variety of documents. In essence, modern emergency management information systems have achieved the single log by combining all of the entries made by the different staff positions in a single journal record.

In large Emergency Operations Centers, with a variety of ongoing actions that have little in common, a single master log may not be practical. In such cases, each major section should maintain its own log.[56]

Although the design of a log form or computer log entry menu may vary widely based on specific needs and past practices, consideration should be given to including the following information:[41,43,236,239]

… date and time of entry
… source (person or organization or communications or sensor media)
… contact information (phone number, radio call sign, e-mail)
… text summarizing the event, information, decision, etc.
… a specific means of identifying whether follow-up activity is required, and, if so, whether it is done
… identity of person making log entry – commonly done by initialing a box or closing the entry with the signature of the staff member on duty.

Maintaining Message Logs

A separate log is commonly used to record all messages received, traditionally as formal written messages transmitted by radio or delivered by a courier. In today's electronic mail environment, this could also include a record of all e-mail messages received. Such message logs are normally maintained in a Message Center, the single focal point in the Emergency Operations Center for incoming communications. The message log is an important means of documenting what messages were received and dispatched by the Emergency Operations Center.

Typically a message log should include the following types of information about each message handled:[53,110]

... whether the message is incoming or outgoing
... the time the message was received in or dispatched from the emergency operations center
... who originated the message
... to whom the message was addressed
... what method was used to transmit the message
... the subject of the message

The Message Center should maintain the message log as a permanent document during the emergency response. A copy of each message handled should be included with the log. This would seem to duplicate the master log maintained in the Emergency Operations Center, but experience suggests that the message log and file may be the only way of quickly retrieving a specific communication that can only be identified by time or subject or originator. It is also vital to being able to prove what communications were received and dispatched in a formal format.

Incident Documentation Completion and Storage

The initial steps toward building a complete file of incident documentation should be taken during the disaster itself. At the completion of each shift, all documents should be checked for completeness, to insure that dates and times are noted, all standard blanks are either completed or marked

N/A (not applicable), individuals responsible have initialed or signed as required, etc. Two important principles apply to this effort: first, if a blank is not filled in, there is a presumption that it was not considered or not done;[23,27] second, although in the heat of the event it may not be possible to complete all paperwork and details, it is generally not acceptable to go back a week or a month later to fill in the details. If additions or corrections to a document are necessary, they must be done as soon as possible after the document was initiated.[89]

At the same time it is important to establish a document file for the incident.[117] One file may be adequate for a single function, although in a major event each position may generate a large volume of documentation, and require a new file each day or for each operational period. All documents handled by a staff function should be placed in the file. No document dealing with the event should leave the Emergency Operations Center at the end of the day - those that leave are almost invariably misplaced, filed in some other file, disposed of, or otherwise lost to the disaster record. It is a good practice to page number the disaster record for each day in order from first to last so that a missing page or document can easily be detected.

With the increasing use of emergency management software, the message control and logging functions are handled to a much greater extent electronically. At the end of set periods established in standard operating procedures, the complete incident log should be printed and filed as a paper copy. This ensures the log against the possibility that a power failure will deny access to or erase the record or that human error will consign it to some undiscoverable destination in the memory.

If the Emergency Operations Center uses status boards as a standard way for displaying information for use by decision makers, the contents of those boards may be important in establishing what information was available as the basis for decision. While most Emergency Operations Centers db not have sufficient personnel to assign someone to maintain a record of what is posted on status boards, some options exist. Individuals who post information can make a second posting to a paper copy of the board. The use of ExlErate software can simplify record keeping of status board contents if a Center also uses Pase's standard status boards. However, a record copy of any status board can be improvised with a word-processing program such as Word or WordPerfect. For critical information it may be worth installing electronic whiteboards that allow either a printed copy to be made of the board or that produce a digital record that can be exported to a computer.[147] However, both of these options are expensive and may not be practical if budgets are limited.

On completion of a disaster response, it is time to convert the event records from a working file into one that is in inactive storage.[207] This should be done only after all response and recovery actions have been completed, including completion of disaster assistance programs, the preparation of any disaster mitigation grants that will be based on hazards identified in this disaster, and the incorporation of lessons learned in training programs.

Completion of event files includes printing copies of all computer records as paper copies. Given the speed of evolution of computer software and hardware, today=s computer discs using today=s software may not be retrievable in three or four years.

As a final step in preparing incident files for storage, the contents of the file should be catalogued. Preparing a list of file contents meets two important objectives: simplifying location of specific types of documents in case of need, and establishing that the files are complete if required in future litigation. As a minimum, a file catalog should list each independent document, a brief description of the document, its length, and its location (for example, "incident log 1 May 1997, 32 pages, file folder 4, box 2").

Incident documents are legal records of what an agency, or a jurisdiction or a business, has done to respond to an emergency. Therefore, they must be retained at least for the period during which the organization or its personnel may be subject to litigation for actions taken during the emergency. This is not as simple as it may sound - in some cases litigation may be initiated based on the point in time at which an injury was discovered, not necessarily when it took place. An agency=s records also may be subject to subpoena for a court case to which the agency itself is not a party. Therefore, policy on how long records should be stored for before being disposed of should be set only after consulting with an attorney. Destruction of records without an established policy, and in a time period that is less than the end of legal liability, creates the impression of cover-up.

In general, public agency disaster documents are open to public inspection under the Freedom of Information Act. Although individual privacy concerns and law enforcement issues may restrict access to some portions of the record, members of the public and the media may request access to the disaster files. If such requests are made, they generally must be honored within time limits set by law and without attempts to evade or capriciously deny access.[270] However, a government agency generally may charge a reasonable fee for reproduction and handling and set reasonable times during which public inspection is permitted.

Disaster records are important historical records. Even if, by law, they may be disposed of, they should not be destroyed. Donation to a local historical museum or society or to a university with an emergency management or disaster studies program should be considered if they are no longer needed for official purposes, and if they do not contain proprietary business information or information that might violate individual privacy.

12. PUBLIC INFORMATION

Media Strategy

Information Release in an Emergency

The news media represent an extremely important resource for effective management of the crisis portion of emergency events. The relationship between the emergency manager and the news media offers the potential for effectively communicating information that the public needs to know, information that will reduce vulnerability, and information that will make management of the crisis easier and more effective.[35,127] To make this work, certain basic principles must be followed:

(1) Positive relationship. Establish a positive relationship with the local and regional news media before there is a disaster. Meet the key staff, discuss their needs, and provide them with press kits that communicate basic information about the emergency management program.[253]

(2) Public Information Officer. Appoint and train an organizational Public Information Officer.[22] Many jurisdictions and organizations already have public information staff members; these individuals should be sent to emergency public information training.

(3) Plan. Include planning for public information releases as part of incident planning. Identify what information needs to be made available to the public, what outcomes are desired from that information release, and when releases will be made.

(4)	Policies for release.	Establish clear policies on who can release information, and brief all personnel. Generally, all of the staff and emergency responders should know to refer all requests for information to the Public Information Officer and be able to direct reporters to that individual's location. This is not to cover up or deny information to the press; it is to assure than one consistent message is communicated based on an understanding of the complete incident.

(5)	Privacy.	Establish clear policies on the release of information that has the potential to intrude on individuals' privacy. For example, it is reasonable to release the names of individuals killed or injured in a disaster – it may not be reasonable to release their names prior to making a good faith effort to notify the next of kin.

(6)	Freedom of information.	For governmental organizations or private organizations operating under government direction, know the contents of the federal and state Freedom of Information Act. In general, the press has a legal right to government information in whatever form that information is currently available. There are limits to this, including privacy of medical records and information required for criminal investigations, but these limits are complicated. Failing to meet reasonable requests may create liability for either legal judgments or administrative disciplinary actions.[270]

(7)	Be forthcoming and truthful.	Answer press questions honestly.[196,274] Provide as complete an answer as possible, but always be ready to say "we do not have that information at this time" if you do not know the answer.[112] Do not lie, cover up errors or failures, or make up information to avoid appearing uninformed. Not only is this unethical, but those who do will be found out in the long run and, at the very least, will face strong public criticism.[81]

(8)   Be cooperative. Press needs can generally be integrated into response activities if an effort is made. However, do not compromise the safety of disaster responders or members of the press. If it is unsafe, explain that it is unsafe and why and provide the best possible safe access.

(9)   Be timely. An initial news conference or release should be made within the first hour of the response. Try to accommodate press deadlines for stories[127] and to provide information at times that will ensure best access to the public. Information released at 2100 may be in time for the 2300 television news, but it will miss the majority of the possible viewing audience.

(10)   Push information. Regular press releases and press briefings on an announced schedule are preferable. This eliminates a certain amount of pressure on the Public Information Officer to respond to individual requests and builds confidence on the part of the media that the operation is giving them the information they need.[264]

(11)   Suppress rumors. Disasters generate a tremendous amount of incorrect information. Never allow incorrect information to remain unchallenged. If an interviewer asks a question that includes a rumor or that implies an incorrect interpretation of an event, correct it in a straight forward, nonconfrontational, and factual way that makes it clear the public needs correct information. Not all reporters are experts in major emergency response, and facts

that are obvious to emergency managers may be far from clear to them.[274]

(12)　No speculation. Do not speculate or offer criticism of others. It may be easy to second guess the actions of other response agencies and very tempting to want to establish a personal position with the media as an expert commentator. However, speculation without all of the facts can create additional rumors and unfairly damage the reputation of individuals and agencies.[22] It is unethical and unprofessional to publicly criticize the performance of others, especially during an ongoing event or investigation.

(13)　Stay within agency limits. The news media will ask any spokesperson to comment on any facet of an emergency response. Restrict remarks to areas in which the organization had a role in the response. Do not answer questions about another organization's portion of the event, even if the spokesperson thinks he or she knows the answer. Refer those questions to representatives of that organization.

(14)　Do not give exclusives or play favorites. Every member of the press deserves access to the same information, especially if that information is intended to preserve public safety. Selectively releasing information to one type of media or one outlet in preference to others ensures that those slighted will be far less likely to treat your organization favorably in future events.[22]

Types of News Media

News media can be described based on the medium they use to communicate information and on their location. Each influences your media strategy.

The media now includes organizations that regularly disseminate information in at least five ways:[22,253]

(1) Daily print media. The daily newspaper remains a staple of how people in the United States receive news. Although many cities were once served by morning and evening newspapers, most markets are now served by a single newspaper, usually a morning paper. The daily newspaper traditionally offers the most in-depth reporting, and usually has news staff with expertise in both the community and its issues and in particular areas of staff member interest. The daily newspaper remains a viable way of providing general public education information and event specific information in relatively slowly developing disasters.

(2) Periodic print media. Smaller communities may still be served by a weekly newspaper, usually with a distinctively local focus. Weekly newspapers are of relatively little utility in developing events, but they offer an excellent forum for local human interest stories and public education materials tailored to specific community needs.

(3) Radio. Radio news offers the most rapid way to disseminate emergency instructions and information on rapidly developing disaster conditions in the community. In many communities, certain stations are well recognized by the public as sources for emergency information in severe weather. The Emergency Alerting System (formerly the Emergency Broadcast System) allows direct access by emergency management for the broadcasting of time sensitive emergency information.

(4)   Television. Television news operates around the cycle of morning show, noontime news, early evening news, and late evening news, with the largest viewing audience in most areas being the early evening news. In addition, some stations will provide short updates and even dedicated coverage of major disaster events as the event is happening. Television provides visual as well as audio coverage of disasters and can shape public perceptions about both the impact of and response to an emergency. Television offers detailed coverage of weather events that is not available in any other format and that may complement or hinder governmental efforts to deal with a weather disaster. Unlike other media, television offers significant opportunities for gathering data on disaster impacts through broadcast images.

(5)   Internet based media. The impact of the Internet as a news source is beginning to emerge. Most major newspapers now post their key stories on Internet pages, and major Internet Service Providers offer regularly updated news stories as a part of their service. The ease of establishing an Internet site means that sites will emerge during disasters to post a wide variety of information, much of it unconfirmed rumor, and some of it conspiracy theories of the event. This has the potential to significantly complicate public information efforts.

At the same time coverage of events on organizational web sites has the potential to help control rumors, disseminate official information to wide audiences, make press releases to the media, and significantly reduce telephone inquiries. Use of a web site as an information tool embedded in the Emergency Operations Center provides the organization an opportunity to tell the correct story without slant or filter – in essence the Emergency Operations Center becomes a member of the media. Its value also extends beyond public access to serving in an internal information role for other units of the organization or jurisdiction.[70,205]

In addition, the media may be classed as national, regional, or local based on their geographical area of focus and their marketing. For example, in television, the national networks and their news are national in focus, as is the case with The Weather Channel and Cable News Network. However, individual affiliate stations of the networks retain a regional or local interest. Among newspapers, only USA Today, the New York Times, and the Washington Post are really national in their coverage and perspective. Some newspapers achieve state or regional coverage – in general the newspaper of the state capitol becomes the newspaper of record for the state. From an emergency management perspective, the more local in focus a newspaper is, the more likely its coverage will be of value in disseminating local emergency information and the more interested the newspaper is likely to be in a local jurisdiction story.[274]

Media Information Requirements

Although it is possible to generalize, each media outlet has its own, often unique, information requirements. If emergency public information efforts are to be effective, the Public Information Officer must know the following:

(1) Deadlines. It takes time to transform information into a story that appears in print the next morning or on the 6:00 news. Know those deadlines.[253] In general, television and radio news has a chance of getting on the air as late as an hour before broadcast time; most

morning newspapers are completed in the early evening. For some types of public education information or meeting announcements, this deadline may be two weeks in advance.

(2)   Access.   News departments have their own telephone and fax lines and even e-mail addresses. Having the telephone numbers is critical when the main switchboard shuts down or goes to recorded messages at the end of the work day.[81]

(3)   When the news department shuts down.   If critical information needs to be disseminated at 10:30 pm, do not select an outlet with a news department that goes home for the night at 6:00 pm.

(4)   Know the format.   Most print media reporters will work from a press release, followed up by a telephone or on-site interview if they are interested in the topic.   Radio may conduct a telephone interview.   Television stories often will involve an interview of a department spokesperson.

(5)   Pictures.   Television is a visual medium; this is also true of newspapers, although to a lesser extent. Television and newspaper reporters will want to get video and still footage of disaster scenes, interviews, emergency equipment, etc.

(6)   Human interest.   All media like stories that involve their readers; one way to do that is to interview either individual emergency workers or disaster victims.   It is appropriate to make individual responders available to talk about their part of the disaster.   It is helpful if the individuals interviewed have been trained in basic interview techniques and are positive spokespeople for the organization.   There is

debate about whether or not the media should be denied access to disaster victims; the trend is clearly in the direction of isolating aircraft crash victim families from the media, for example. However, over-managing victims may well backfire if the victims perceive they are being managed, again with examples of this obvious in recent air disasters.

Internal Public Information

Most large organizations have internal public information activities that inform the employees, volunteers, and other organizational stakeholders. Any public information strategy must include the use of available tools such as the organizational newsletter, e-mail distribution of information bulletins, and phone message broadcast systems to make current, correct information available. This helps to control rumors, improves work performance, increases the effectiveness of impact preparations, and improves morale by providing clear guidance.

Press Releases

The most basic tool for communication with the media is the prepared press release. Even media that rely heavily on interviews will use the press release as a starting point in preparing the story.

Press Release Format

Using the commonly accepted press release format organizes the information for the news reporter and often results in a press release being printed as written. News stories try to answer a key series of questions; think of them as five Ws and a H. WHAT happened? WHERE did it happen? WHEN did it happen? WHY did it happen? WHO was involved? HOW did it happen? News stories also are measured by the column inch. What is printed is often a reflection of the balance of stories versus space. That means that well written stories go from the most important to the least important. The first paragraph has the most important information, and the last paragraph the least. If space limits mean something gets cut, the cut is made counting back from the end, and the portion cut will always be less important than the portion printed. General principles for preparing a press release include:[22,253,270]

(1)   Letterhead. Use your standard stationery or a specific press release form that identifies the release clearly as a press release.

(2)   Date. Date the release. Press releases often include a line FOR IMMEDIATE RELEASE just below or above the date; this is somewhat redundant, as it would not seem to make sense to send out releases that are not intended for immediate release.

(3)   Contact information. At the top of the page include the name, telephone number, and e-mail address of the organization point of contact (normally the Public Information Officer).

(4)   Spacing and margins. Allow a minimum of one inch margins and double space the body of the release.

(5)   Headline. Lead with a short headline that describes the topic of the release. Often this headline will not be used, but it serves as a subject line to attract the attention of the news staff.[20]

(6)   Short and to the point. Write the article using simple language and short complete sentences. Avoid use of acronyms and specialist language with which the general public will be unfamiliar. In general, completed releases that will be read on radio or television should be one page in length and releases for print media one or two pages.

(7)   More. If your release runs more than one page in length, center MORE at the bottom of each page except the last. Start each page after the first with the topic, your organization name, and the page number.

(8)   End. At the bottom of the last page center ### END ###.

Prepared Press Releases

The Public Information Officer should be prepared to issue an initial press release very early in a developing disaster, especially if the news media will be used to disseminate warnings and provide instructions to the public. One of the best ways to do this is to have standard news releases prepared for the types of disaster events you can reasonably expect. These can be fill-in-the-blanks documents that can be rapidly prepared on a word processor by completing time, location, impact area, shelters opened, and similar data.[22] The completed word processing file can then be faxed using computer fax software to a list of media newsroom fax numbers and posted on the agency's web site.

Interviews

Any emergency manager can expect at some point in a crisis to be interviewed by the press. Interviews are inherently difficult situations. Managers are under the stress of managing a difficult event, often with either too much information or incomplete and limited information, and their focus is on the problems the impact is causing. A lucky emergency manager will have had minimal training in interview skills and been interviewed once or twice in a career. He or she has no idea what the reporter's personal agenda is. And, given normal disaster work patterns, he or she is probably exhausted. The individual doing the interviewing knows the agenda, has done hundreds of interviews, knows what questions to ask and what responses they want from the manager, and ultimately controls what he or she is reported as having said. This means that emergency managers must follow some basic rules to ensure their message is the one that is reported.[81]

(1)   Look the part. Most people will understand if a uniform shows the effects of fighting the fire or flood. But an individual who look like he or she was dirty to start with, wearing a uniform that is nonprofessional or a caricature or inappropriate for conditions, and who appears rattled and unsure, will loose all credibility. Spend the extra minute to attend to grooming, straighten up clothing, and control emotions and fatigue before the interview starts.

(2)   Stay on the message. Enter an interview with a message the public needs to hear. Focus on that message, take every opportunity to restate it, and do not be distracted from it.[81,196]

(3)   Think before speaking. Compose the message carefully, with a logical order, and using words that the audience will understand. Remember that the general public will not understand specialized emergency response

218

terminology and that even common (to emergency responders) acronyms will be unfamiliar to them.[196] Remember also that the interview may be fed to the national media and seen almost anywhere in the world. Speak clearly and consciously avoid slang. Do not make jokes to relieve tension.[112,253] It goes without saying that funny remarks about people in other states or in other countries will not be funny to them and that most people will not be impressed by racist, sexist, or other discriminatory statements.

(4)   Take the time needed. It is not the interviewee's responsibility to fill air time for the reporter. Pause, analyze the question, think through the response, and then speak. There is no obligation to give rapid, poorly conceived answers to questions.

(5)   Stop when finished.[253] We are uncomfortable with silence in our conversations. Interviewers use this discomfort creatively by waiting after the speaker has finished speaking in the hopes that the silence will cause him or her to make additional, poorly thought through comments. These comments may take the manager off the intended message and create opportunities for the interviewer to probe for weakness or inconsistency in the comments.

(6)   Have facts and figures.[81] Every response generates a wide variety of facts and figures – how many people are injured, how high the water is, the number of responding units, the exact time of the earthquake, the number and locations of shelters opened, etc. Have that data available on a pad for reference, and use these facts throughout the interview. A calm, authoritative statement of facts is hard to argue with and conveys a clear impression that the speaker knows what is happening.

(7)   Answer the real question.   Not all questions will be straightforward.   Be prepared for questions that (a) rephrase the message in a significantly different way as part of searching for clarification or verification[175] to send the reporter's message, not the speaker's, (b) include rumors or incorrect information or ask "what if",[172] (c) are designed to pry out information not yet ready for release,[174] (d) contain false premises,[173] or (e) are phrased so that any answer makes the speaker look foolish.   In these cases, do not hesitate to correct errors or rumors and tell the message the way it should be told, not to answer the booby-trapped question that was asked.   Remember that viewers will remember the answer, not the question the reporter asked.[81]

(8)   Off the record.   Nothing said to a reporter is off the record or for deep background – anything said may show up in print with a name attached to it.[81,112,196]   Some authors indicate in some limited situations it is possible to go off the record to obtain media agreement to not release critical information in such events as a terrorist attack.[270]   However, this is a dangerous tactic.   Always assume anything said, even after completion of an interview, can be picked up by a microphone and will be used by the reporter.[253]

(9)   Never say "no comment."   No comment is an admission of guilt, and waves the red flag of cover-up.[253]   If the speaker has no comment, refer the reporters to the Joint Information Center, indicate that the subject is outside the speaker's area of knowledge or expertise, state that no information is available on the subject yet, or whatever other response is accurate and truthful.[81,274]

(10)   Never criticize the victims.   Even if it is obvious that people died because they did things that were incredibly

stupid, it is bad form to say they were idiots.[274] It is much better to leave discussions of causation to the investigation phase. "The pilot crashed the airplane because he was stupid" (an actual statement made by a Public Information Officer and reported in the press in a search and rescue incident) really upsets the pilot's surviving relatives and makes officials seem incredibly heartless.

(11) Do not ask for a copy of the story to review prior to its publication. This immediately raises suspicions of censorship of media content.[274]

(12) Agency policy. Some organizations have a policy that requires that all press contacts be reported – typically including which media, reporter's name, subjects discussed (and especially subjects that elicited strong reporter interest), and expected publication time. Immediately after the interview, make notes so the contact report will be accurate – it is a good practice to ask the reporter for a business card, but without referring to the reason (this also adds to your database of local press contacts).

Public Information Functions

Public Information Officer

The Public Information Officer is the staff officer responsible for managing an effective public information program for a specific disaster event. This includes:[1,28,48]

(1) Developing public information tactics for the event, advising the staff managing the incident, and addressing public information concerns in daily planning,

(2)   Collecting and processing current incident information to provide an accurate summary for dissemination,

(3)   Managing the flow of information to the various media, including releasing press releases, conducting press conferences, and arranging for interviews,

(4)   Monitoring what media outlets are saying about the event so that errors can be corrected and positive messages reinforced,[35]

(5)   Setting up any appropriate public information facilities, and

(6)   Managing public information unit staffing and work flow.

To fulfill these responsibilities, the Public Information Officer requires regular access to the staff, the incident site, and the leaders in charge of the response. The Public Information Officer must have the confidence of, and direct access to, the chief elected official of the jurisdiction, senior executive leadership, and the emergency manager and incident commanders.[81]

Public Information Staffing

In a large event, the workload will exceed the ability of a single Public Information Officer to manage the responsibilities listed above. If staffing is available, the Public Information Officer function may be expanded to provide staff responsible for:

(1)   Information gathering, including at incident Command Posts and through monitoring the media.

(2)   Information dissemination, with a rumor control unit, media briefers, a staff member responsible for the Emergency Alerting System, and a staff member managing internal information dissemination.

(3)   Field operations, with staff members to gather information and validate or exclude rumors, provide assistance to the media, perform community liaison, and document operations with still and video photography.[62]

Joint Information Center

During major events, the potential exists for each agency to release a different version of the event, causing confusion among the general public. To reduce the likelihood of conflict among public information efforts, Joint Information Centers are now a standard feature of emergency response. The Joint Information Center becomes the single point of interface with the press, and the participating agencies speak with a single voice[112,264] through a designated lead public information officer.[22] This ensures the public receives a single, coordinated message that addresses the concerns of all participating organizations. It provides the press with a single location for press releases, news conferences and interviews, access to officials, and support services. A subsidiary benefit to all organizations is that it allows pooling of public information resources to provide adequate 24 hour staffing. The Joint Information Center is normally established in proximity to the Emergency Operations Center or the Command Post, but not as a part of the actual command and control facility.

Rumor Control

Any major event will generate a significant amount of misinformation. A percentage of this is the inevitable result of the difficulty in collecting and processing situation information during a large, complex event that disrupts the normal sensor systems and reporting arrangements of a community. The insertion of human participants playing a variety of roles into the disaster environment leads to an equal variety of interpretations of observed events. Eventually, rumors flow, as reports are passed from person to person in a disaster version of an old children's game.

To prevent inappropriate citizen responses and the diversion of resources to deal with shadow problems, rumor control is an important function. Effective rumor control depends on several elements:

(1)   Training. All disaster responders should be trained to identify and contain rumors.

(2)   Good information gathering, processing, and dissemination within the emergency management structure. A sizable percentage of rumors are circulated by the response forces; the key to stopping this is training and making sure the responders have as much accurate information as possible.

(3)   An effective working relationship with the media. Accurate and rapid dissemination by the media of official information can reduce rumors and make the public's response more effective.

(4)   A Rumor Control Center.   All Emergency
Operations Centers and Joint Information Centers should be
supported by a Rumor Control Center.   This call center has
adequate telephone lines and is staffed to answer public
inquiries and provide current, correct information.[81]

(5)   Web site.   An emergency web site with
current information has the potential to significantly reduce
telephone inquiries.   In addition, supporting information,
such as maps, lists of shelters opened, evacuation
instructions, instructions on how to prepare residences and
businesses for the disaster impact, etc. can provide the public
more information in a format they can print and use
rapidly.[70,205]

13. STRESS

Stress and Emergency Responders

The effects of incident stress on emergency responders have been widely discussed, and it is generally accepted that major emergencies and disasters can generate significant stress for participants.[103,146,151,275] Participants in field operations, including in a field Command Post, will be exposed to these effects. Dispatchers in Public Safety Answering Points have also been recognized as a population subject to incident stress.[113] However, it is not generally recognized (in many cases even by the participants) that staffs of Emergency Operations Centers are similarly subject to stress, even though their operational environment is quite similar to that of dispatchers. And, although critical incident stress management programs have been highly touted as the solution for incident stress, the effectiveness of specific programs in dealing with individual issues is open to some question.[8,92,118]

Sources of Stress in the Emergency Operations Center

The physical environment, the nature of the work, and the nature of the event combine to create a stressful environment in an Emergency Operations Center. Factors at work include:

(1) The poor physical design of most Emergency Operations Centers is a common factor. Facilities are typically crowded, with too many workers for the existing air conditioning and sanitary system.[151] Work stations are smaller than workers are accustomed to in their normal jobs. Furniture, especially seating, is often not designed for

sustained use over long periods of time, leading to ergonomic problems and physical fatigue. Noise level is typically high, contributing to fatigue and stress.[13,113]

(2) The typical Emergency Operations Center is cluttered, not only because of limited space, but also because of the high volume of paper documents generated. The clutter generates stress,[151] compounded by frustration resulting from the difficulty of locating needed documents or tracking the status of actions.

(3) Individuals working in the Emergency Operations Center may know their normal duties very well. However, they may now have unfamiliar additional responsibilities, uncertainty as to their decision-making authority, and little training in their agencies' disaster roles. In many cases, their understanding of the functions of other agencies may be limited, leading to conflicts over turf.[184] This is compounded by having to work as part of an unfamiliar structure reporting to supervisors they do not know, and whose authority is also uncertain.[251]

(4) Shift lengths of 12 hours, or even until the individual physically can no longer function, are normal. As a result individuals are fatigued and may be unable to gain sufficient rest during the course of an extended event to ever perform at a normal level. Research suggests the effects of fatigue from a 17 hour period (12 hour shift plus travel and personal time) are equivalent to a 0.5 blood alcohol content, and from 24 hours a 1.0 blood alcohol content.[219] The effects of sleep deprivation may be in addition to the effects of disruption of normal sleep cycle, a demonstrated cause of industrial and highway accidents.[45]

227

(5)   There is a tendency to rely on coffee, pizza, fried chicken, and other available fast foods as the staple of Emergency Operations Center diet.   Although convenient, the nutritional value of such meals is open to question.[113]

(6)   The nature of the work performed in Emergency Operations Centers creates great uncertainty and brings little closure.   Staff members are continually required to make decisions that involve putting resources at risk with minimal information.   Although some operations may be completed during a shift, others will extend for days, providing little sense of successful accomplishment.   And often, the disaster is simply too great for effective control, leading to a perception of failure.   This is compounded by the access of the staff to reports from across the jurisdiction, magnifying perceptions of the degree of the disaster.[184]

(7)   Second guessing by the media, elected officials, the general public, and even members of the emergency services is commonplace.   This is exacerbated by the lack of recognition for the staff's efforts.   The rescue worker who makes a heroic effort gets his or her picture in the newspaper and is honored with medals and certificates.   The staff member in the Emergency Operations Center who correctly assessed the situation, identified the correct resources, and provided the rescuers the information needed to reach the scene and complete the rescue gets no recognition.

(8)   And the Emergency Operations Center staff is separated from reality.   At the simplest level, this means they do not know what is happening outside the center and have little idea what they will face when the shift ends and they walk outside (even at the level of whether it is daylight or dark outside).   However, this can translate into real concerns

about safety of family members and property to add to the stress of managing the incident.[184]

Strategies for Stress Control

Although there are larger numbers of stressors, many of these can be controlled and eliminated if a stress control strategy is adopted.[97]

(1)   Regular training provides staff an understanding of their duties and role in the Emergency Operations Center.[151,275]   It also gives them a chance to work in the Center's organizational structure so that supervisory relationships will not be strange to them, and builds a sense of comradeship among the staff.[113,184]

(2)   Although the physical facilities may be quite limited, careful use of space may make it possible to reduce crowding and give workstations adequate work areas.[146] One of the highest priorities should be spending the money needed to purchase chairs that roll, swivel, and tilt and that provide adequate support for extended sitting.   A second priority should be painting the interior of the facility a bright, cheerful color.

(3)   Put clutter controls in place.   Have filing cabinets, folders, bins, and notebooks available.   Punch all forms for insertion in three ring binders – each position can maintain a complete record of its activities in a single binder, rather than in piles of loose paper in in-baskets.   Make only the number of copies actually needed.   Reduce the number of forms needed by each staff member to the absolute minimum.   And enforce a clean desk policy.

(4)   Hold a regular staff briefing, at least once every two hours.   This ensures everyone is informed, improving performance and reducing uncertainty and stress.

(5)   Do family protection training for the family members of the staff.   Not everyone will take advantage of it, but making basic disaster safety and shelter information available may reduce the stress level of staff members.[184]

(6)   Enforce breaks.[184]   Every hour each staff member should get out of his or her chair and go for a walk, inside the building, out into the parking lot, wherever conditions will permit.

(7)   Push adequate nutrition and hydration.[151]   Consuming diuretics such as coffee, or other caffeine based drinks, complicates stress by dehydrating the staff.   Have healthy food available along with water and juices.   Some individuals will get caught up in the action and forget to take basic care of themselves.   Use of a buddy system to manage breaks, nutrition, and hydration may be useful.

(8)   Control shift lengths.[81,184]   No one should work sustained 12 hour shifts, and absolutely no one should work more than 12 hours.

(9)   Recognize performers.   Emergency Operations Center staff should be recognized for completing training and for superior performance during emergencies.   Involve their normal supervisors and the public information staff in this recognition.   And thank their families in writing or through a family day for their support.

(10)   Critique each event. An after action review covering the event, the actions of the Emergency Operations Center, procedures that worked well, and areas for improvement serves several purposes.   On the one hand it fulfills legitimate operational and training needs.   However, it also allows staff members to place their experiences in context, deal with unresolved issues, and move forward in a positive way.[113]   If possible, this should be scheduled after termination of the event and at a time convenient for maximum attendance.

14. STANDARDS AND ETHICS

Emergency Services and Disaster Laws and Ordinances

State Laws

The most basic expression of standards of performance and conduct is found in the law. State disaster and emergency services laws provide the legal authority for disaster operations within a state. Typically such laws include provisions for:[51,145]

(1) A process for the declaration of a state of emergency or disaster. Such declarations are a formal acknowledgment by a jurisdiction that a major emergency is in progress and that certain actions must be undertaken to effectively manage response to the emergency.

(2) Assignment of legal authority to direct emergency operations to the chief elected officer of the jurisdiction (Governor, Chairman of the County Council, Mayor, etc.) and provision for appointment of a coordinator to manage emergency management programs.

(3) Suspension of certain provisions of state law or administrative regulations if needed to effectively manage the emergency.

(4) Broad protection from liability for individuals performing disaster response duties.

(5) Changes in purchasing authority and personnel regulations to allow needed supplies to be obtained quickly and temporary workers to be hired.

(6) Assignment of authority for coordination of state government response to the state's emergency management agency.

(7) Establishment of a state Emergency Operations Plan as an executive order of the Governor with legal authority to govern state government response.

(8) Provision of authority for the ordering of voluntary and mandatory evacuations and other population protection measures.

Local Ordinances

Local ordinances may be enacted by county, city, and town governments to provide local authority for disaster response.[65,256] The content of such ordinances varies widely, but will generally tend to parallel the provisions of the state disaster laws. In many cases local jurisdiction Emergency Operations Plans are adopted by the elected governing body, giving these documents similar force to that of an ordinance and giving local government authority to direct a wide variety of response actions.

Voluntary Standards

A variety of organizations use a consensus process to develop voluntary standards that may impact organizational response to emergencies. Although these standards are voluntary, they may be incorporated into law by reference, giving them the same force as the rest of the state or federal code. In litigation they may be used to establish the standard of practice that should be followed by a reasonable man.

As consensus documents, standards represent the agenda of the organization that adopts them, not necessarily best, common, or accepted practices in the field. The standards making process generally gives the standards committee wide latitude to reject input that disagrees with the positions of the committee members. Although standards organizations do attempt to achieve some balance in committee membership, it is not unheard of for a standards committee to primarily represent a limited constituency.

The primary voluntary standard for disaster management is <u>NFPA 1600 Standard on Disaster/Emergency Management and Business Continuity Programs</u>, issued by the National Fire Protection Association. NFPA 1600 incorporates the commonly accepted phases of emergency management (mitigation, preparedness, response, and recovery) as a framework for establishing minimum requirements for administration of an emergency management program, but does not provide substantive guidance for crisis operations.[156]

The primary voluntary standard for incident command systems is <u>NFPA 1561 Standard on Fire Department Incident Management System</u>. NFPA 1561 provides broad guidance for the structure of an incident command or management system, but does not mandate any particular model.[155]

Codes of Ethics

Codes of ethics express the moral values of a profession and provide a moral compass to guide members of the profession in their dealings with their organizations, each other, and the public. Because of the unique characteristics of emergency response operations, the maintenance of a high standard of ethical performance is vital to individuals who work in the emergency services, emergency management, and business continuity professions. Effective emergency response depends on confidence in the word and integrity of the people with whom you are working and in whose hands you may place your life or the safety and security of others. This is not a place to be with a liar, cheat, or thief.

Codes of ethics are voluntary standards, usually established by professional bodies, either for the profession at large, or (and more convincingly) for their own members. Although codes of ethics generally do not have enforcement mechanisms, clearly ethical conduct does have benefits. It increases trust in the individual and the organization by others. Adherence to a code of ethics protects individuals and organizations to some degree from claims of unethical conduct. It also should increase public and client satisfaction with the services provide by the organization.

It is important to remember that codes of ethics exist because people have done the types of things that codes of ethics say are inappropriate and the results have been damaging to people and organizations. When the code of ethics for individuals certified as Certified Crisis Operations Managers was developed, the individuals working on the code based most of the standards on actual cases of inappropriate behavior with which they were familiar.

Public Administration Code of Ethics

The code of ethics adopted by the American Society for Public Administration provides a code for the conduct of all public agency employees and volunteers. Because emergency services and emergency management personnel generally fall within this category, this code provides a baseline applicable to all types of activities in these fields.

Figure 14-1. Major Elements of the American Society for Public Administration Code of Ethics[11]

I. Serve the Public Interest: Serve the public, beyond serving oneself.

II. Respect the Constitution and the Law: Respect, support, and study government constitutions and laws that define responsibilities of public agencies, employees, and all citizens.

III. Demonstrate Personal Integrity: Demonstrate the highest standards in all activities to inspire public confidence and trust in public service.

IV. Promote Ethical Organizations: Strengthen organizational capabilities to apply ethics, efficiency and effectiveness in serving the public.

V. Strive for Professional Excellence: Strengthen individual capabilities and encourage the professional development of others.

Emergency Management Code of Ethics

The code of ethics adopted by the International Association of Emergency Managers (Table 14-2) provides a code tailored to emergency management.

Figure 14-2. International Association of Emergency
Managers Code of Ethics and Professional Conduct[116]

PREAMBLE:

Maintenance of public trust and confidence is central to the
effectiveness of the Emergency Management Profession. The
members of the International Association of Emergency
Managers (IAEM) adhere to the highest standards of ethical
and professional conduct. This Code of Ethics for the IAEM
members reflects the spirit and proper conduct dictated by
the conscience of society and commitment to the well-being
of all. The members abide by the Association's core values of
RESPECT, COMMITMENT, and PROFESSIONALISM.

VALUES:

RESPECT

Respect for supervising officials, colleagues, associates, and
most importantly, for the people we serve is the standard for
IAEM members. We comply with all laws and regulations
applicable to our purpose and position, and responsibly and
impartially apply them to all concerned. We respect fiscal
resources by evaluating organizational decisions to provide
the best service or product at a minimal cost without
sacrificing quality.

COMMITMENT

IAEM members commit themselves to promoting decisions
that engender trust and those we serve. We commit to
continuous improvement by fairly administering the affairs
of our positions, by fostering honest and trustworthy

relationships, and by striving for impeccable accuracy and clarity in what we say or write. We commit to enhancing stewardship of resources and the caliber of service we deliver while striving to improve the quality of life in the community we serve.

PROFESSIONALISM

IAEM is an organization that actively promotes professionalism to ensure public confidence in Emergency Management. Our reputations are built on the faithful discharge of our duties. Our professionalism is founded on Education, Safety and Protection of Life and Property.

Business Continuity Code of Ethics

The code of ethics developed by The Disaster Recovery Institute International (see Table 14-3) provides a code of ethics for business continuity professionals. It is important to note that business ethical standards legitimately differ from those for employees of public organizations and institutions. What may be inappropriate conduct for a public employee may be perfectly reasonable and proper for an employee of a business or corporation.

The CCOM Code of Ethics

Certification as a Certified Crisis Operations Manager (CCOM) carries with it an expectation that certified individuals will behave in an ethical manner. The CCOM Code of Ethics (see Table 14-4) expresses these expectations in a formal way. Certified individuals are expected to know and abide by the provisions of this document in their professional roles as emergency responders, emergency managers, or business continuity professionals. Failure to do so may result in withdrawal of the Certification.

Figure 14-3. Code of Ethics for Business Continuity Professionals[71]

As certified business continuity professionals, we will:

Practice the highest level of professionalism at all times in the performance of our duties.

Practice conduct that is legal and ethical and will avoid any perception of conflict of interest for ourselves, our employers, and our clients.

Practice and promote corporate continuity and disaster recovery concepts.

Keep confidential any information revealed as such in the performance of our duties.

Continually seek to increase our competence and the competence of those who work with us.

Participate in continuing professional knowledge and skill improvement programs.

Figure 14-4. The Certified Crisis Operations Manager Code
of Ethics

As a Certified Crisis Operations Manager I will:

Protect life, property, the environment, and the economic and social stability of my community from effects of disasters.

Assure the safety of the response workers performing duties under my direction.

Perform my duties fairly without the appearance or actuality of partisanship, favoritism, or conflict of interest.

Treat colleagues and constituents with dignity and respect.

Tell the truth in my dealings with others.

Represent my qualifications accurately and realistically assess and portray the capabilities of my organization.

Protect emergency resources from theft, waste, and abuse.

Protect confidential and proprietary information and the privacy of disaster victims from inappropriate disclosure.

Continue to develop myself at every opportunity through professional reading, training, and emergency exercises.

Participate actively in the profession through emergency management and business continuity organizations.

Strive to achieve the highest degree of excellence in the performance of my duties.

WORKS CITED

This list of works cited was developed using ProCite 5, a bibliography and citation generating software. The style selected, Uniform, was chosen because the use of number citations as superscript appeared to be the least disruptive for those using the volume as originally intended in preparing for the Certified Crisis Operations Manager examination.

1. Field Operations Guide ICS-420-1. Stillwater, OK: Fire Protection Publications, 1983.

2. Operations Corner. Community Alert Network, Inc. Newsletter. Summer 1995; 6-7.

3. Who's the Wise Guy? A Message about Sheltering. NCCEM Bulletin. Jul 1995; 12(7): 4.

4. Front cover. 9-1-1 Magazine. Sep–Oct 1995; 8(4).

5. Traveling Cell. 9-1-1 Magazine. Sep-Oct 1995; 8(4): 8.

6. Interstate E.M. Compact Gets Congressional Okay. NCCEM Bulletin. Nov 1996; 13(11): 2.

7. SVI Custom Innovations for Out-of-Space Fire Departments. Fire Rescue Magazine. Mar 2000; 19(3): 15.

8. Research questions suitability of debriefing for civilian staff. Civil Protection. Apr 2000; (49): 15.

9. New Deliveries. Fire Rescue Magazine. Mar 2001; 19(3): 75.

10. Bellingham Emergency Siren System [Web Page]. Available at http://ps01.upd.wwu.edu/sirentest.htm. (Accessed May 27, 2001).

11. American Society for Public Administration. ASPA=s Code of Ethics [Web Page]. Available at http://www.aspanet.org/member/coe.htm. (Accessed May 11, 2000).

12. American Society for Testing and Materials. F 1288-90 Standard Guide for Planning for and Response to a Multiple Casualty Incident. Philadelphia, PA: ASTM, 1990.

13. Anderson DJ. Ready Or Not ... Here Comes Disaster! Lenhartsville, PA: Summit House Publishers, 1978.

14. Anderson PB. Rural Major EMS Incident Scene Management Manual. Lincoln, NE: Lincoln Medical Education Foundation, n.d.

15. Antal JF. Armor Attacks: The Tank Platoon. Novato, CA: Presidio Press, 1991.

16. Arnold G. Refuges of Last Resort [Web Page]. Nov 17, 1997; Available at http://fepa.org/fepa_init_refuges.htm. (Accessed May 27, 2001)

17. Atkinson W. The Next New Madrid Earthquake: A Survival Guide for the Midwest. Carbondale, IL: Southern Illinois University Press, 1989.

18. Auf der Heide E. Disaster Response: Principles of Preparation and Coordination. St. Louis, MO: C. V. Mosby Company, 1989.

19. Aurick L. Packaged Emergency Station. QST. Sep 1992: 105.

20. Beaudoin T. Bad P.R. 101. Management Review. Dec 1988; 77(12): 44-8.

21. Bellamy S. Car 269: Fireground Radio. Popular Communications. May 1994; 12(9): 28-9.

22. Bernstein AB. The Emergency Public Relations Manual. Third ed. Highland Park, NJ: PASE Incorporated, 1987.

23. Bevelacqua AS. Prehospital Documentation: A Systematic Approach. Englewood Cliffs, NJ: Prentice-Hall, Inc., 1992.

24. Bingham RC. Improving Fireground Radio Communications. Fire Engineering. Feb 1997; 150(2): 38-49.

25. Bolt BA. Earthquakes. New York, NY: W. H. Freeman and Company, 1993.

26. Bowman L. New hurricane response: 'Hide from the wind; flee the flood' [Web Page]. Jul 1, 2000; Available at http://www.naplesnews.com/hurricane/00/d279875a.htm. (Accessed May 27, 2001).

27. Brown-Nixon C. Documentation: The Credibility Skill. Lake Worth, FL: EES Publications, 1989.

28. Brunacini AV. Fire Command. Quincy, MA: National Fire Protection Association, 1985.

29. Cahill J. New Year's Eve in Times Square. Journal of Emergency Medical Services. Dec 1997; 22(12): 38-45.

30. California. Governor's Office of Emergency Services. Standardized Emergency Management System Guidelines: Part I System Description. Sacramento, CA: Governor's Office of Emergency Services, 1995.

31. California. Governor's Office of Emergency Services. Auxiliary Communications Service. Southern Region. ACS Mission [Web Page]. May 19, 2001; Available at http://www.races.net/sca. (Accessed May 26, 2001).

32. Capon RS. A Fishing Tackle HF Station To Go! QST. Nov 1994: 67-8.

33. Carlson EP, Ed. Incident Command System. Stillwater, OK: Fire Protection Publications, 1983.

34. Carter S. Citizen's Band "Portable Base Station." Alexandria, VA: Virginia Defense Force, n.d.

35. Cashman JR. Tips From The Hazardous Materials Pros. n.p.: John R. Cashman, 1983.

36. Christen HT, Maniscalco PM. The EMS Incident Management System: EMS Operations for Mass Casualty and High Impact Incidents. Upper Saddle River, NJ: Prentice-Hall, Inc., 1998.

37. City of Fredericksburg (Virginia). Emergency Operations Center. Message Form. Fredericksburg, VA: City of Fredericksburg, n.d.

38. City of Hampton (Virginia). Office of Emergency Management. Commonwealth of Virginia and City of Hampton Readiness Postures [Web Page]. Mar 8, 2001; Available at http://www.hampton.va.us/eoc/hamready.html. (Accessed May 25, 2001).

39. City of Raymond (New Hampshire). Office of Emergency Management. Activation [Web Page]. Available at http://users.rcn.com/raymondoem/Activation.htm. (Accessed May 25, 2001 May).

40. Civil Air Patrol. Alaska Wing. Tabular Reports: Volume I, Standard Tabs. Anchorage, AK: Alaska Wing Civil Air Patrol, 1988.

41. Civil Air Patrol. Alaska Wing. Wing Operations Center Form 6. Anchorage, AK: Alaska Wing Civil Air Patrol, 1990.

42. Civil Air Patrol. Colorado Wing. Aviation Risk Assessment. Denver, CO: Colorado Wing Civil Air Patrol, 1996.

43. Civil Air Patrol. Colorado Wing. Group 3 Incident Command Team. Incident Log. Colorado Springs, CO: Group 3 Incident Command Team, 1991.

44. Clawson JJ, Dernocoeur KB. Principles of Emergency Medical Dispatch. Englewood Cliffs, NJ: Prentice Hall, 1988.

45. Coburn E. Shiftworker Fatigue: The $77 Billion Problem [Web Page]. 1997; Available at http://www.alertness.com/cost_engineerig.htm. (Accessed May 18, 2001).

46. Coile RC. California's Standardized Emergency Management System. Journal of the American Society of Professional Emergency Planners 1996: 37-9.

47. Coleman JE. Accountability and the Incident Management System. Fire Engineering. Dec 1997; 150(12): 56-9.

48. Coleman JF. Incident Management for the Street-Smart Fire Officer. Saddle Brook, NJ: PennWell Publishing Company, 1997.

49. Coleman JF. "To-Do" Lists Enhance Incident Management. Fire Engineering. Mar 1997; 150(3): 111-24.

50. Colorado. Department of Public Safety. Division of Disaster Emergency Services. Message Form. Golden, CO: Colorado Division of Disaster Emergency Services, n.d.

51. Commonwealth of Virginia. Virginia State Fire, EMS and Emergency Related Laws. 1999/2000 ed. Charlottesville, VA: Michie, 1999.

52. Commonwealth of Virginia. Department of Military Affairs. Virginia Defense Force. Staff Directive Form. Sandston, VA: Virginia Defense Force, 1994.

53. Commonwealth of Virginia. Department of Military Affairs. Virginia Defense Force. Virginia Defense Force Communications Center Message Log. Sandston, VA: Virginia Defense Force, 1995.

54. Commonwealth of Virginia. Governor's Task Force on Emergency Medical Response Disaster Planning. Report of the Governor's Task Force on Emergency Medical Response Disaster Planning. Richmond, VA: Virginia Office of Emergency Medical Services, 1988.

55. Commonwealth of Virginia. Statewide Mutual Aid Committee. Statewide Mutual Aid Implementation Guidebook. Richmond, VA: Virginia Department of Emergency Management, 2001.

56. Commonwealth of Virginia. Virginia State Area Command. Virginia State Area Command Emergency Operations Center Standard Operating Procedures. Richmond, VA: Virginia State Area Command, 1994.

57. Compton D. Risk Versus Gain ... What's the Plan? Speaking of Fire. Oct 1997 Oct; New Series 5(2): 7 and 21.

58. Cook JL. Standard Operating Procedures and Guidelines. Saddle Brook, NJ: Fire Engineering Books and Videos, 1998.

59. Cook JL. Writing Standard Operating Procedures and Guidelines. Fire Engineering. Aug 1999; 152(8): 107-16.

60. Cook S. HTML-Based PrePlans. FireRescue Magazine. Apr 2001; 19(4): 61-5.

61. Cornell AH. The Decision-Maker's Handbook. Englewood Cliffs, NJ: Prentice-Hall, Inc., 1980.

62. Cowan F. Operations, Plans, Logistics, Finance and ... EPI? 9-1-1 Magazine. Mar-Apr 1998; 11(2): 50-1.

63. Cowardin DH. New Radio Codes Create Confusion. American Fire Journal. Aug 1998; 50(8): 11.

64. Crystal Wind Communciations. If A Siren Sounds [Web Page]. Available at http://onyx.xtalwind.net/cc/fpnucl/nuke3.htm. (Accessed May 27, 2001).

65. Daines GE. Planning, training, and exercising. In: Drabek TE, Hoetmer GJ. Emergency Management: Principles and Practice for Local Government. Washington, DC: International City Management Association, 1991: 161-200.

66. Davis L. Environmental Disasters: A Chronicle of Individual, Industrial, and Governmental Carelessness. New York, NY: Facts on File, Inc., 1998.

67. Davis TH. The Patrol Order. Vienna, GA: Old Mountain Press, Inc., 1994.

68. dePolo CM, Rigby JG, Johnson GL, Jacobson SL, Anderson JG. Planning Scenario for a Major Earthquake in Western Nevada. Reno, NV: Nevada Bureau of Mines and Geology, 1996.

69. Dialogic Communications. the communicator! Franklin, TN: Dialogic Communications, 2000.

70. DiMartino K, Evans J. Two Experiences with the Internet: Ft. Collins Disaster Recovery. NCCEM Bulletin. Jan 1998; 15(1): 13-4.

71. Disaster Recovery Institute International. Code of Ethics for Business Continuity Professionals [Web Page]. Available at http://www.dr.org/general.html. (Accessed May 11, 2000).

72. Ditzel P. Fire Alarm! The Fascinating Story Behind The Red Box On The Corner. New Albany, IN: Fire Buff House Publishers, 1990.

73. Drabek TE, Tamminga HL, Kilijanek TS, Adams CR. Managing Multiorganizational Emergency Responses: Emergent Search and Rescue Networks in Natural Disaster and Remote Area Settings. Boulder, CO: University of Colorado Institute of Behavioral Science, 1981.

74. Duncan MJ, Green WG, Kahn IA, Player MB. Virginia Mass Casualty Incident Management Module I: Awareness Level. Richmond, VA: Virginia Office of Emergency Medical Services, 2000.

75. Dunn V. Command and Control of Fires and Emergencies. Saddle Brook, NJ: Fire Engineering Books and Videos, 1999.

76. Dworsky PI. Disaster Communications: It's Not Just Radios. IAEM Bulletin. Dec 1999; 16(12): 7.

77. Edwards JE. Combat Service Support Guide. 2nd ed. Harrisburg, PA: Stackpole Books, 1993.

78. El Paso County (Colorado). General Message. Colorado Springs, CO: El Paso County, 1989.

79. Emergency Management Telecommunications. Reverse 911. Orlando, FL: EMTEL, Inc., 2000.

80. Fagel MJ. Creating a Disaster Plan. IAEM Bulletin. Jul 2000; 17(7): 7-8.

81. Fink S. Crisis Management: Planning for the Inevitable. New York, NY: American Management Association, 1986.

82. FIRESCOPE California. Fire Service Field Operations Guide ICS 420-1. n.p.: FIRESCOPE California, 1996.

83. Flin R. Sitting in the Hot Seat: Leaders and Teams for Critical Incident Management. Chichester, West Sussex, United Kingdom: John Wiley & Sons Ltd., 1996.

84. Florida. Department of Community Affairs. Division of Emergency Management. Statewide Mutual Aid Agreement for Catastrophic Disaster Response and Recovery. Tallahassee, FL: Florida Division of Emergency Management, 1994.

85. Florida. Department of Community Affairs. Division of Emergency Management. Situation Report No. 1: Hurricane Floyd [Web Page]. Sep 12, 1999; Available at http://www.dca.state.fl.us/eoc/EOC%20 Activations/Hurricane%20Floyd/S0912-01.html. (Accessed May 27, 2001).

86. Florida. Department of Community Affairs. Division of Emergency Management. State Emergency Operations Center Activation Levels [Web Page].

Available at http://www.dca.state.fl.us/eoc/
eoclevel.htm. (Accessed May 25, 2001).

87. Florida Fire Chiefs' Association. Statewide Fire-
Rescue Disaster Response Plan. Ormond Beach, FL:
Florida Fire Chiefs' Association, 1997.

88. Fradkin PL. Magnitude 8: Earthquakes and Life
Along the San Andreas Fault. New York, NY: Henry
Holt and Company, 1998.

89. Frew SA. Street Law: Rights and Responsibilities of
the EMT. Reston, VA: Reston Publishing Company,
Inc., 1983.

90. Garnham P. Airway Management. Emergency. Jul
1997; 29(7): 12-5.

91. Gillespie DF. Coordinating community resources. In:
Drabek TE, Hoetmer GJ. Emergency Management:
Principles and Practice for Local Government.
Washington, DC: International City Management
Association, 1991: 55 78.

92. Gist R, Lohr J, Kenardy J, Bergmann L, Meldrum L,
Redburn B, et al. Researchers Speak on CISM.
Journal of Emergency Medical Services. May 1997;
22(5): 27-8.

93. Gove PB. Webester's Third New International
Dictionary of the English Language Unabridged.
Springfield, MA: Merriam-Webster, Inc., 1993.

94. Gray C. No Way Out - Roads can't handle
evacuation; local escape routes would flood or jam

[Web Page]. Jun 1, 1998; Available at http://www. nolalive.com/weather/articles_99/tpw060198b_nowa yout.html. (Accessed May 27, 2001).

95. Green WG. The Virtual Emergency Operations Center. At the 2000 Conference of the State and Local Emergency Management Data Users Group. Orlando, FL: 2000.

96. Green WG. Virginia's Volunteer Program to Staff a Medical Emergency Operations Center. Journal of the American Society of Professional Emergency Planners. 1997: 32-4.

97. Green WG. Get Ready To Manage Stress. Emergency Manager. Winter 1997; (1): 12.

98. Green WG. E-Emergency Management: the Operational State of the Art. In: Kowalski KM, Trevits MA. Contingencies, Emergency, Crisis, and Disaster Management: Emergency Management in the Third Millenium: 7th Annual Conference of The International Emergency Management Society. Orlando, FL: The International Emergency Management Society, 2000.

99. Green WG. Emergency Management Acronyms. Parkland, FL: Universal Publishers, 2001.

100. Green WG. The State of Local Government Emergency Operations Centers: The Impact on Computer Based Emergency Management. At the 2001 Conference of the State and Local Emergency Management Data Users Group. Las Vegas, NV: 2001.

101. Green WG, Player MB, Schwartz TJ, Adams MP, Glover RL, Edwards BW, et al. Virginia Mass Casualty Incident Management Module II: Operations Level. Richmond, VA: Virginia Office of Emergency Medical Services, 1999.

102. Hadfield P. Sixty Seconds That Will Change The World: The Coming Tokyo Earthquake. Boston, MA: Charles E. Tuttle Company, 1991.

103. Hafen BQ, Frandsen K. Psychological Emergencies and Crisis Intervention: A Comprehensive Guide for Emergency Personnel. Englewood Cliffs, NJ: Prentice-Hall, Inc., 1985.

104. Hansen DW. Amateur Radio Survival Kits. QST. Oct 1995: 91.

105. Hansen W. The Alaska Earthquake, March 27, 1964: Effects on Communities. Washington, DC: U. S. Government Printing Office, 1965.

106. Hartin F. Tactical Decision Making: Games We Should Play, Part 1. Fire-Rescue Magazine. Jan 1998; 16(1): 84-8.

107. Hausman M, Scace S. Voices In The Sky: Satellite Communications and Disaster Response. 9-1-1 Magazine. Mar-Apr 1998; 11(2): 24-7.

108. Heppenheimer TA. The Coming Quake: Science and Trembling on the California Earthquake Frontier. New York, NY: Times Books, 1988.

109. Higgins JM. 101 Creative Problem Solving

Techniques: The Handbook of New Ideas for Business. Winter Park, FL: The New Management Publishing Company, 1994.

110. High Country REACT. Message Center Traffic Log. Colorado Springs, CO: High Country REACT, 1991.

111. Hoetmer GJ. Introduction. In: Drabek TE, Hoetmer GJ. Emergency Management: Principles and Practice for Local Government. Washington, DC: International City Management Association, 1991: xvii-xxxiv.

112. Hoffman C. An Interview on Crisis Communications. NCCEM Bulletin. Jan 1998; 15(1): 9-10.

113. Holt FX. The Top 10 Things You Should Know About Dispatcher Stress. 9-1-1 Magazine. Nov-Dec 1997; 10(6): 18-25.

114. Hospital Emergency Incident Command System Update Project. HEICS III [Web Page]. Available at http://www.emsa.ca.gov/dms2/heics3.htm. (Accessed May 24, 2001).

115. Hurder L. Public Safety Communications Manual. Newington, CT: American Radio Relay League, Inc., 1990.

116. International Association of Emergency Managers. IAEM Code of Ethics and Professional Conduct [Web Page]. Available at http://www.iacm.org/About_IAEM/body_ethics.html. (Accessed May 11, 2000).

117. International Association of Fire Chiefs. Fire Service Emergency Management Handbook. Washington, DC: U. S. Government Printing Office, 1985.

118. Jacobson B. Is CISM Always Necessary? 9-1-1 Magazine. Nov-Dec 1997; 10(6): 20-2.

119. Jakubowski G. In Command: Run the Fireground from one location. Fire-Rescue Magazine. Jul 2000; 18(7): 98.

120. Jenaway WF. Pre-Emergency Planning. Ashland, MA: International Society of Fire Service Instructors, 1986.

121. Jones B. Building the Camper's Portable Hamshack. QST. Apr 1995; 60-1.

122. Kandel JI. In Andrew's Wake. QST. Dec 1992; 76(12): 20-8.

123. Kellogg A. Emergency Operations Centers For Public Health. IAEM Bulletin. Jul 1999; 16(7): 14 5

124. Kipp JD, Loflin ME. Emergency Incident Risk Management. New York, NY: Van Nostrand Reinhold, 1996.

125. Kramer WM, Bahme CW. Fire Officer's Guide to Disaster Control. Saddle Brook, NJ: Fire Engineering Books & Videos, 1992.

126. Landesman LY. Emergency Preparedness in Health Care Organizations. Oakbrook Terrace, IL: Joint Commission on Accreditation of Healthcare

Organizations, 1996.

127. Lankford DM. Working with the Media During a Crisis. NCCEM Bulletin. Jan 1998; 15(1): 6-7.

128. Larson E. Isaac's Storm: A Man, a Time, and the Deadliest Hurricane in History. New York, NY: Crown Publishers, 1999.

129. Larson R. Branch Mountain Relay: Wildfire Communications Behind the Scenes. 9-1-1 Magazine. Mar-Apr 1997; 10(2): 43-7.

130. Larson RD. Mobile Command Posts. 9-1-1 Magazine. Sep-Oct 1995; 8(4): 38-47.

131. Larson RD. New Directions in Disaster Mapping Systems. 9-1-1 Magazine. Jan-Feb 1997; 10(1): 46-7.

132. Larton D. Incident Dispatcher Teams. 9-1-1 Magazine. Mar-Apr 1997; 10(2): 46.

133. Larton D. Return of the Wild Ones: Communications Pre-Planning for Hollister Biker Festival. 9-1-1 Magazine. Nov-Dec 1997; 10(6): 26-30.

134. Larton D. On-Line Weather Resources for the PSAP. 9-1-1 Magazine. Mar-Apr 1998; 11(2): 19-21.

135. LeSage P. Fire & Rescue Field Guide. Lake Oswego, OR: InforMed, 1996.

136. Levy M, Salvadori M. Why The Earth Quakes. New York, NY: W. W. Norton Company, Inc., 1995.

137. Lewin R, McFadden J. Online Incident Command.

Fire-Rescue Magazine. Oct 1997; 15(8): 54-7.

138. Longest BB, Rakich JS, Darr K. Managing Health
Services Organizations and Systems. Baltimore, MD:
Health Professions Press, Inc., 2000.

139. Longshore D. Encyclopedia of Hurricanes,
Typhoons, and Cyclones. New York, NY:
Checkmark Books, 2000.

140. Lunsford DS. Organizing an EOC on a Shoestring
Budget. IAEM Bulletin. Jul 1999; 16(7): 6-8.

141. Martin D. Three Mile Island: Prologue or Epilogue.
Cambridge, MA: Ballinger Publishing Company,
1980.

142. McCoy LC. Crisis Management as a Function of
Information Exchange. NCCEM Bulletin. Jan 1998;
15(1): 15-8.

143. McMillian JR. An Aural Brevity Code for Public
Safety Communications. New Smyrna Beach, FL:
Associated Public-Safety Communications Officers,
Inc., 1974.

144. Medvedev G. The Truth About Chernobyl. n.p.:
Basic Books, Inc., 1991.

145. Michigan. Department of State Police. Emergency
Management Division. Michigan Emergency
Management Act 1990 (Act 390 of 1976, as
Amended). Lansing, MI: Michigan Department of
State Police, 1990.

146. Mitchell JT, Resnik HLP. Emergency Response to Crisis. Bowie, MD: Robert J. Brady Company, 1981.

147. Moore P. Corporate Emergency Operations Centers. IAEM Bulletin. Jul 1999; 16(7): 9-10.

148. Moore P. How to Expand Emergency Response Plans to Address Recovery and Continuity of Operations. IAEM Bulletin. Feb 2001; 18(2): 12-3.

149. Moore W. Developing an Emergency Operations Center. Washington, DC: International City/County Management Association, 1998. (IQ Service Report; vol 30).

150. Murname J. Document This ... Journal of Emergency Medical Services. Jan 2000; 25(1): 48-54.

151. Myers D. Worker Stress During Longterm Disaster Recovery Efforts. In: Everly GS. Innovations in Disaster and Trauma Psychology. Vol. One. Ellicott City, MD: Chevron Publishing Corporation, 1995: 158-91.

152. Nance JJ. On Shaky Ground: An Invitation to Disaster. New York, NY: William Morrow and Company, Inc., 1988.

153. National Association for Search and Rescue. Incident Commander Field Handbook: Search and Rescue. Fairfax, VA: National Association for Search and Rescue, 1987.

154. National Fire Protection Association. Report of Committee on Fire Service Occupational Safety and

Health: NFPA 1561 Standard on Fire Department Incident Management System. Quincy, MA: National Fire Protection Association, 1989.

155. National Fire Protection Association. NFPA 1561 Standard on Fire Department Incident Management System. Quincy, MA: National Fire Protection Association, 1995.

156. National Fire Protection Association. NFPA 1600 Standard on Disaster/Emergency Management and Business Continuity Programs. Quincy, MA: National Fire Protection Association, 2000.

157. National Fire Service Incident Management System Consortium Model Procedures Committee. Model Procedures Guide for Structural Firefighting. Stillwater, OK: Fire Protection Publications, 1993.

158. National Fire Service Incident Management System Consortium Model Procedures Committee. Model Procedures Guide for Emergency Medical Incidents. Stillwater, OK: Fire Protection Publications, 1996.

159. National Interagency Incident Management System. Incident Command System Position Manual: Finance Section. Stillwater, OK: Fire Protection Publications, n.d.

160. National Interagency Incident Management System. Incident Command System Position Manual: Logistics Section. Stillwater, OK: Fire Protection Publications, n.d.

161. National Interagency Incident Management System.

Incident Command System Position Manual: Operations Section. Stillwater, OK: Fire Protection Publications, n.d.

162. National Interagency Incident Management System. Incident Command System Position Manual: Planning Section. Stillwater, OK: Fire Protection Publications, n.d.

163. National Interagency Incident Management System. Incident Command System Position Manual: Command Section. Stillwater, OK: Fire Protection Publications, n.d.

164. National Wildfire Coordinating Group. Area Command: Module 15, I-400, Instructor Guide. Boise, ID: National Interagency Fire Center, 1994.

165. National Wildfire Coordinating Group. ICS Orientation: Module 1, I-100, Instructor Guide. Boise, ID: National Interagency Fire Center, 1994.

166. National Wildfire Coordinating Group. Incident and Event Planning: Module 11, I-300, Instructor Guide. Boise, ID: National Interagency Fire Center, 1994.

167. National Wildfire Coordinating Group. Multi-Agency Coordination: Module 16, I-401, Instructor Guide. Boise, ID: Boise Interagency Fire Center, 1994.

168. National Wildfire Coordinating Group. Unified Command: Module 13, I-400, Instructor Guide. Boise, ID: National Interagency Fire Center, 1994.

169. National Wildfire Coordinating Group. Wildland Fire

Qualification Subsystem Guide. Boise, ID: National
Interagency Fire Center, 1999.

170. Nielson D. The Utah County SCATeam. CQ VHF.
Jul 1996; 1(6): 69-72.

171. Oahu (Hawaii). Civil Defense Agency. Tsunami
General Information [Web Page]. Available at
http://www.co.honolulu.hi.us/ocda/guide11.htm.
(Accessed May 27, 2001).

172. Onder J. Responding to Questions in a Hostile Media
Environment. Responder Magazine. Sep 1997; 4(9):
18.

173. Onder J. Responding to Questions in a Hostile Media
Environment. Responder Magazine. Nov 1997;
4(11): 14.

174. Onder J. Responding to Questions in a Hostile Media
Environment. Responder Magazine. Dec 1997; 4(12):
12.

175. Onder J. Responding to Questions in a Hostile Media
Environment. Responder Magazine. Jan 1998; 5(1):
22.

176. Pagonis WG, Cruikshank JL. Moving Mountains:
Lessons in Leadership and Logistics from the Gulf
War. Boston, MA: Harvard Business School Press,
1992.

177. Palm Beach County (Florida). Division of
Emergency Management. Regional Evacuation
Coordination Procedures: Refuges of Last Resort

Plan [Web Page]. Aug 2000; Available at
http://www.co.palm-beach.fl.us/EOC/Revised
%20Draft%20Refuge%20of%20Last%20Resort%20
Plan%20.htm. (Accessed May 27, 2001).

178. Palm R. ARES Field Resource Manual. Newington,
CT: American Radio Relay League, 1995.

179. Pase Incorporated. Pase Incorporated Products ...
[Web Page]. Available at http://www.paseinc.com/
products.html. (Accessed May 28, 2001).

180. Perrow C. Normal Accidents: Living with High-Risk
Technologies. Princeton, NJ: Princeton University
Press, 1999.

181. Perry DG. Wildland Firefighting: Fire Behavior,
Tactics & Command. Bellflower, CA: Fire
Publications, Inc., 1987.

182. Perry DG. Managing A Wildland Fire: A Practical
Perspective. Bellflower, CA: Fire Publications, Inc.,
1989.

183. Perry LG. Preparing for an Emergency: A Step-by-
Step Approach. In: Disaster Recovery World III. St.
Louis, MO: Disaster Recovery Journal, 1996.

184. Perry RW. "Managing disaster response operations".
In: Drabek TE, Hoetmer GJ, Eds. Emergency
Management: Principles and Practice for Local
Government. Washington, DC: International City
Management Association, 1991: 201-23.

185. Phelps BW. Incident Command System Policy &

Procedures Manual. MD: Phelps Publishing Company, 1991.

186. Pickett G. Planning for Sudden Hazardous Events: Tactical Considerations for Mass Casualty Incidents, School Shootings. IAEM Bulletin. Jul 2000; 17(7): 6.

187. Pioveson B. A Guide To Emergency Net Operations [Web Page]. Available at http://home.xnet.com/~w9ib/ecom.html. (Accessed May 26, 2001).

188. Pivetta S. 9-1-1 Emergency Communications Manual. Auburn, WA: Professional Pride, 1991.

189. Procops T. Buyer Beware: The Inside Wire on Call Logging Systems. 9-1-1 Magazine. Jan-Feb 2001; 14(1): 38-42.

190. Purtle D. The Tech Side: Staying In Contact During A Crisis. NCCEM Bulletin. Jan 1998; 15(1): 11-2.

191. Quade ES. Analysis for Public Decisions. New York, NY: Elsevier Science Publishing Company, Inc., 1982.

192. REACT International Inc. Team Contact Directory. Wichita, KS: REACT International, Inc., 1992.

193. Reinhardt J. A Van for All Reasons. QST. Feb 1995: 40-2.

194. Ronshagen C. AVL Is Coming of Age. 9-1-1 Magazine. Jan-Feb 2001; 14(1): 30-1.

195. Root D. Expanding Your Horizons: When The Creek

Becomes A Raging River. 9-1-1 Magazine. Mar-Apr 1998; 11(2): 60-1.

196. Rosenbaum M. How to Conquer Mediaphobia. Management Review. Dec 1988; 77(12): 41-3.

197. Rosenthal U, Charles MT, Hart PT. Coping With Crises: The Management of Disasters, Riots, and Terrorism. Springfield, IL: Charles C. Thomas, Publisher, 1989.

198. Rubin DL, Maniscalco PM. "The EMS Safety Officer: Incident Accountability". Emergency Medical Services. Apr 2001; 30(4): 52-5.

199. Sachs GM. Multiple Casualty Incident Management, Part I. Fire Engineering. Dec 1997; 150(12): 73-5.

200. Salvation Army Team Emergency Radio Network. SATERN [Web Page]. Available at http://home.xnet. com/~w9ib/satern.html. (Accessed May 26, 2001).

201. San Joaquin County (California). Office of Emergency Services. Damage Summary Status Board [Web Page]. Mar 11, 1997; Available at http://www. co.san-joaquin.ca.us/oes/disasters/12-96storms/ 97HTML.htm. (Accessed May 28, 2001).

202. Saniter DJ. Phased Decision-Making. Journal of the American Society of Professional Emergency Planners. 1998; 85-91.

203. Schaper FC. Accountability Simplified - The Dispatcher's Role: Part II. 9-1-1 Magazine. Nov-Dec 1997; 10(6): 96-7.

204. Schaper FC. Incident Command: Managing The Large Scale Incident. 9-1-1 Magazine. Mar-Apr 1998; 11(2): 44-7.

205. Schwert DP. Two Experiences with the Internet: In Response to the 1997 Flood on the Red River of the North. NCCEM Bulletin. Jan 1998; 15(1): 13-6.

206. Sensenig DD. Creating an Emergency Management Support Team For Extraordinary EOC Staffing Needs. NCCEM Bulletin. Aug 1997; 14(8): 8.

207. Setnicka TJ. Wilderness Search and Rescue. Boston, MA: Appalachian Mountain Club, 1980.

208. Sibley BJ. Coming Soon ... A "Virtual" EOC? IAEM Bulletin. Jul 1999; 16(7): 13.

209. Simmons T. Phoenix's high-tech dispatch center. Fire Engineering. Jul 1997; 150(7): 86-92.

210. Smith JE. Developing a Mutual Aid Attitude in the Dispatch Center. 9-1-1 Magazine. Mar-Apr 1997; 10(2): 40-2.

211. Smith JE. Mutual Aid In Cyberspace. 9-1-1 Magazine. Jan-Feb 1998; 11(1): 18-22.

212. Southern Governor's Association. Southern Regional Emergency Management Compact. Washington, DC: Southern Governor's Association, 1993.

213. Spencer RE. The Evolution of an EOC - From Fallout-Proof Bunker to High-Tech Center. IAEM Bulletin. Jul 1999; 16(7): 11-2.

214. Stabler D. ICS in the Communications Center. 9-1-1 Magazine. Mar-Apr 1998; 11(2): 48-9.

215. Stafford County (Virginia). Emergency Operations Center. Message Form. Stafford, VA: Stafford County, n.d.

216. Steinberg M, Potter G. Weather Alert Systems for the PSAP. 9-1-1 Magazine. Mar-Apr 1998; 11(2): 18-22.

217. Stevens LH. It's Up to Our Officers. Fire-Rescue Magazine. May 1997; 15(3): 8.

218. Streger MR. Mass Casualty and Disaster Communications. Emergency Medical Services. Apr 1999; 28(4): 59-62.

219. Swann P. New countermeasures to reduce young driver accidents [Web Page]. Available at http://www.dot.wa.gov.au/roadsafety/papers/new_countermeasurers.html. (Accessed May 18, 2001).

220. Tazelaar E. Those Wonderful RACES Volunteers. IAEM Bulletin. Dec 1999; 16(12): 6-8.

221. TeleMinder. TeleMinder Emergency Management. Los Altos, CA: TeleMinder, 2000.

222. Terry RJ. Organizing a Contingency Plan for Success. In: Disaster Recovery World III. St. Louis, MO: Disaster Recovery Journal, 1996.

223. Texas. Department of Public Safety. Division of Emergency Management. San Bernard River Hazardous Materials Incident: SITUATION

REPORT # 2. Austin, TX: Texas Division of
Emergency Management, 2001.

224. The Associated Public-Safety Communications
Officers, Inc. The Public Safety Communications
Standard Operating Procedure Manual. 22 ed. South
Daytona, FL: The Associated Public-Safety
Communications Officers, Inc., 1990.

225. The Information Company. Chem/Bio Survival Cards
for Civilians. n.p.: The Information Company, 1999.

226. The Virtual Emergency Operations Center. Exercise
HURREX 2001 SITREP 3 [Web Page]. Apr 4, 2001;
Available at http://www.virtualeoc.org/eoc/
sitrep.html. (Accessed May 27, 2001).

227. Tscherne RF. The Do-It-Yourself Ready Made
Warning Order & Operation Order. n.p.: Richard F.
Tscherne, n.d.

228. United Kingdom. British Information Services. Front
Line: The Official Story of the Civil Defense of
Britain. New York, NY: The MacMillan Company,
1943.

229. United States Army. United States Army Infantry
School. Ranger Training Brigade. Ranger Handbook.
Fort Benning, GA: U. S. Army Ranger School, 1992.

230. United States. Department of Agriculture. Forest
Service. Cooperative Fire Protection. National
Interagency Incident Management System. Boise, ID:
Boise Interagency Fire Center, 1983.

231. United States. Department of Commerce. National Oceanic and Atmospheric Administration. National Weather Service. tornadoes ... Nature's Most Violent Storms. Washington, DC: U. S. Government Printing Office, 1992.

232. United States. Department of Commerce. National Oceanic and Atmospheric Administration. National Weather Service. thunderstorms and lightning ... the underrated killers. Washington, DC: U. S. Government Printing Office, 1994.

233. United States. Department of Commerce. National Oceanic and Atmospheric Administration. National Weather Service. hurricanes ... Unleashing Nature's Fury. Washington, DC: U. S. Government Printing Office, 1996.

234. United States. Department of the Air Force. USAF Operational Planning Process. Washington, DC: U. S. Government Printing Office, 1986.

235. United States. Department of the Air Force. Chief Air Force MARS. United States Air Force Military Affiliate Radio System [Web Page]. May 2000; Available at http://public.afca.scott.af.mil/public/mars1.htm. (Accessed May 26, 2001).

236. United States. Department of the Army. Daily Staff Journal or Duty Officer's Log. Washington, DC: U. S. Government Printing Office, 1962.

237. United States. Department of the Army. FM 101-5 Staff Organization and Operations. Washington, DC: U. S. Government Printing Office, 1984.

238. United States. Department of the Army. FM 7-71 Light Infantry Company. Washington, DC: U. S. Government Printing Office, 1987.

239. United States. Department of the Army. Madigan Army Medical Center. EOC Event Log [Web Page]. Available at http://fmace.amedd.army.mil/fepp/ Secix.html. (Accessed May 28, 2001).

240. United States. Department of the Army. Madigan Army Medical Center. Personnel Status Board - Management Team [Web Page]. Available at http://fmace.amedd.army.mil/fepp/Status_board_mgt. htm. (Accessed May 28, 2001).

241. United States. Department of the Interior. Bureau of Land Management. Oregon State Office. Response to Continuity of Operations Plan [Web Page]. Jun 12, 1998; Available at http://www.or.blm.gov/efoia/fy98/ IMs/m98068.htm. (Accessed May 27, 2001).

242. United States. Department of Transportation. Coast Guard. U. S. Coast Guard Auxiliary Air Operations Manual. Washington, DC: U. S. Government Printing Office, 1981.

243. United States. Department of Transportation. United States Coast Guard. Radiotelephone Communications. Washington, DC: U. S. Government Printing Office, 1983.

244. United States. Federal Communications Commission. Emergency Alert System 2001 AM & FM Handbook. Washington, DC: Federal Communications Commission, 2001.

245. United States. Federal Emergency Management Agency. Emergency Operating Centers Handbook. Washington, DC: U. S. Government Printing Office, 1984.

246. United States. Federal Emergency Management Agency. A Guide for the Review of State and Local Emergency Operations Plans. Washington, DC: U. S. Government Printing Office, 1992.

247. United States. Federal Emergency Management Agency. Urban Search and Rescue Response System Field Operations Guide. Revised ed. Washington, DC: U. S. Government Printing Office, 1993.

248. United States. Federal Emergency Management Agency. Guide for All-Hazard Emergency Operations Planning. Washington, DC: U. S. Government Printing Office, 1996.

249. United States. Federal Emergency Management Agency. Federal Response Plan. 9230.1-PL. Washington, DC: U. S. Government Printing Office, 1999.

250. United States. Federal Emergency Management Agency. Emergency Management Institute. The California FIRESCOPE Program. Washington, DC: U. S. Government Printing Office, 1987.

251. United States. Federal Emergency Management Agency. Emergency Management Institute. Public Policy in Emergency Management: Student Manual. Washington, DC: U. S. Government Printing Office, 1990.

252. United States. Federal Emergency Management
   Agency. Emergency Management Institute. State
   and Local Continuity of Government (COG): Student
   Manual. Washington, DC: U. S. Government Printing
   Office, 1990.

253. United States. Federal Emergency Management
   Agency. Emergency Management Institute. Basic
   Public Information Course: Student Manual.
   Washington, DC: U. S. Government Printing Office,
   1994.

254. United States. Federal Emergency Management
   Agency. Emergency Management Institute. EOC's
   Management and Operations Course: Student
   Manual. Washington, DC: U. S. Government Printing
   Office, 1995.

255. United States. Federal Emergency Management
   Agency. Emergency Management Institute. Disaster-
   Related Needs of Seniors and Persons with
   Disabilities: Student Manual. Washington, DC: U. S.
   Government Printing Office, 1997

256. United States. Federal Emergency Management
   Agency. Emergency Management Institute. The
   Emergency Program Manager. Emmitsburg, MD:
   Emergency Management Institute, n.d.

257. United States. Federal Emergency Management
   Agency. United States Fire Administration. National
   Fire Academy. Incident Command System for
   Emergency Medical Services: Student Manual.
   Washington, DC: U. S. Government Printing Office,
   1995.

258. United States. Federal Emergency Management Agency. United States Fire Administration. National Fire Academy. Incident Command System Self Study Unit. Washington, DC: U. S. Government Printing Office, 1995.

259. United States. Federal Emergency Management Agency. United States Fire Administration. National Fire Academy. United States. Department of Justice. Office of Justice Programs. Emergency Response to Terrorism Job Aid. Washington, DC: U. S. Government Printing Office, 2000.

260. United States. Fish and Wildlife Service. 039 FW 1 Continuity of Operations [Web Page]. Sep 17, 1999; Available at http://policy.fws.gov/039fw1.html. (Accessed May 27, 2001).

261. United States. Forest Service. GACC Detailed Situation Report 05/26/2001: Northern California [Web Page]. May 26, 2001; Available at http://www.fs.fed.us/r5/fire/north/sitreport.htm. (Accessed May 27, 2001).

262. United States. National Communications System. Government Emergency Telecommunications Service. GETS Pocket User Guide. Arlington, VA: GETS Program Management Office, n.d.

263. Upland Fire Department. Multi-Casualty ICS. Stillwater, OK: Fire Protection Publications, n.d.

264. Van Hiel KP, Hoo R. Media Case History: Dunblane, Scotland. NCCEM Bulletin. Jan 1998; 15(1): 8-9.

265. Vatter MT. The Impact of Staffing Levels and Fire Severity on Injuries. Fire Engineering. Aug 1999; 152(8): 125-32.

266. Vaughan D. The Challenger Launch Decision: Risky Technology, Culture, and Deviance at NASA. Chicago, IL: The University of Chicago Press, 1996.

267. Virginia Association of Volunteer Rescue Squads, Inc. Disaster Manual. Richmond, VA: Virginia Association of Volunteer Rescue Squads, Inc., 1993.

268. Wade NM. The Battle Staff SMARTbook. Lakeland, FL: The Lightning Press, 1999.

269. Walker B. Planet Earth: Earthquake. Alexandria, VA: Time-Life Books, 1982.

270. Warner J, Sweatman-Ridgeway B. Law Enforcement Media Relations Handbook. Gaithersburg, MD: Jack Warner and Beverly Sweatman-Ridgeway, n.d.

271. Wieder MA. Shifting Gears: How to Make the Right Choice During Incident Operations. American Fire Journal. Mar 2001; 53(3): 12-4.

272. Williams D. Disaster Recovery Planning. In: Disaster Recovery Journal Disaster Recovery World. St. Louis, MO: Disaster Recovery Journal, 1991: 18-28.

273. Wood D. Attack Warning Red: The Royal Observer Corps and the Defence of Britain 1925 to 1975. London, United Kingdom: Macdonalds and Jane's Publishers Limited, 1976.

274. Zetlin M. Meet The Press - And Survive! Management Review. Dec 1988; 77(1): 35-40.

275. Zusman J. Meeting Mental Health Needs in a Disaster: A Public Health View. In: Parad HJ, Resnick HLP, Parad LG. Emergency and Disaster Management. Bowie, MD: The Charles Press Publishers, Inc., 1976:245-58.

INDEX